BestMasters

Springer awards „BestMasters" to the best master's theses which have been completed at renowned universities in Germany, Austria, and Switzerland.

The studies received highest marks and were recommended for publication by supervisors. They address current issues from various fields of research in natural sciences, psychology, technology, and economics.

The series addresses practitioners as well as scientists and, in particular, offers guidance for early stage researchers.

Jochen Szangolies

Testing Quantum Contextuality

The Problem of Compatibility

Springer Spektrum

Jochen Szangolies
Siegen, Germany

BestMasters
ISBN 978-3-658-09199-6   ISBN 978-3-658-09200-9 (eBook)
DOI 10.1007/978-3-658-09200-9

Library of Congress Control Number: 2015933641

Springer Spektrum

Printed on acid-free paper

Springer Spektrum is a brand of Springer Fachmedien Wiesbaden
Springer Fachmedien Wiesbaden is part of Springer Science+Business Media
(www.springer.com)

To my mother, Barbara, in loving memory.

*"You were a presence full of light upon this Earth
and I am a witness to your life and to its worth."*
—John Darnielle, Matthew 25:21

Foreword

A central insight of modern quantum physics is that randomness in the quantum regime has a different nature than in the classical world. In classical theories such as statistical mechanics randomness is explained by missing knowledge on the initial conditions or physical processes. In the quantum regime, however, this view can become problematic. If randomness is caused by ignorance, one assumes that there are additional parameters which determine the process and the randomness disappears for observers knowing these parameters. John Bell, however, showed in 1964 that models with hidden parameters must be non-local, if they should reproduce the predictions of quantum mechanics. The corresponding violation of Bell's inequalities has been observed experimentally, ruling out certain types of hidden variable models.

There is a second no-go theorem for hidden variable models, which is known as the Kochen Specker theorem. In short, it states that quantum mechanics cannot be reconciled with classical models that are noncontextual for compatible observables. Here, compatible observables are observables that can be measured simultaneously or in any order without disturbance, and noncontextuality means that the value of an observable does not depend on which other compatible observable is measured jointly with it. The phenomenon that quantum theory goes beyond noncontextual theories is called quantum contextuality and it also has been recently observed with trapped ions, polarized photons or nuclear magnetic resonance.

When testing quantum contextuality in experiments, a sequence of several measurements is carried out on a single quantum system. Here, a problem arises: the observables should not disturb each other, but due to experimental imperfections this will not be the case. So one has to analyze the effects of this errors on the conclusions drawn from the experimental data and this is exactly what the current thesis from Jochen Szangolies is about. In this thesis, several types of noise models for experiments are discussed. In principle, already small amounts of noise can nullify the conclusions about a contextuality experiment. But, as shown in this thesis, with the help of modified Kochen Specker inequalities one can still rule out large classes of hidden variable models. The proposed modification can easily be implemented, as it only requires measuring permutations of the quantum observables. Therefore, the thesis of Jochen Szangolies represents a valuable contribution to current discussions and I am happy that it appears in the Springer BestMasters book series.

Otfried Gühne

Profile of the Institute

THE Department of Physics at the University of Siegen focuses its research on particle physics, quantum optics and solid state physics. The research of the Theoretical Quantum Optics group deals with three topics: First, the group works on the characterization of quantum entanglement. This concerns the detection and quantification of entanglement, mainly in the multiparticle setting. Second, the group is doing research on the foundations of quantum mechanics. Here, mainly quantum contextuality and the theory of sequential quantum measurements are considered. Finally, the group analyses statistical aspects of quantum experiments. This includes direct collaborations with experimental groups on the analysis and interpretation of their data.

Preface

QUANTUM theory, as formulated in the first three decades of the last century, has long since developed an almost proverbial reputation for being weird, counterintuitive, or even flat-out impossible to understand. The source of this is certainly to be found in the ways in which it differs from classical physics, which until the advent of quantum theory seemed to carry the promise of bringing the world within our grasp—that is, providing a clear and intelligible picture of the mechanisms according to which planets and atoms alike trace their orbits. It is the upset of such intuitions, honed by our intimate familiarity with the macroscopic, that lends its mysterious air to quantum theory.

This book constitutes neither an attempt at dispelling the mysteries, nor does it try to reconcile them with pre-quantum expectations. Rather, its aim is to contribute to the more modest programme of making precise where and how quantum physics diverges from classical expectations. The—albeit reluctant—founding fathers of said programme are Albert Einstein, Boris Podolsky, and Nathan Rosen, who in their seminal article "Can Quantum-Mechanical Description of Physical Reality be Considered Complete?" [2] were the first to raise the issue of the completeness of quantum theory, that is, the question of whether it provides a full account of the underlying physics, or whether it has to be augmented by additional quantities (which have since become known as 'hidden variables') to furnish such a description.

Their intent was, with an ingenious argument, to establish that there indeed are quantities that quantum mechanics in its standard form does not account for, and thus, to show that there must be some deeper theory capable of describing this more fundamental layer.

They could not have anticipated that their argument would ultimately, via the mediation of John Bell, lead to very stringent constraints on the possibility of such a more fundamental description: as Bell showed [7], if we are to entertain the possibility of definite quantities left unaccounted for by the quantum formalism, then these quantities must be able to instantaneously influence one another across arbitrary distances, if the resulting theory is to be capable of duplicating all quantum mechanical predictions. The experimental confirmation of these predictions then was what cemented the importance of Bell's theorem in the corpus of scientific knowledge.

The present work is concerned mainly with a related, though substantially different theoretical result, typically known as the Kochen-Specker theorem after Simon Kochen and Ernst Specker [23]. Like Bell's, the theorem by Kochen and Specker considers the completability of quantum mechanics with additional quantities, and derives constraints on the nature of these quantities. Unlike Bell's theorem, experimental assessment of its consequences is less straightforward. The problem is, in a nutshell, that the relevant notion of noncontextuality—roughly, the persistence of physical quantities independently of the experimental setting—is well defined only for ideal measurements, and ceases to apply in a more realistic setting. The resulting obstruction to the experimental testability of the Kochen-Specker theorem is the main topic of this monograph.

The work contained herein was accepted as a diploma thesis at the University of Siegen, and was completed under the tutelage of Prof. Dr. Otfried Gühne, for whose guidance, experience, and not least for the opportunity of working within his group on this fascinating topic I could not be more grateful. An equal measure of thanks is due to Dr. Matthias Kleinmann, who took me under his wing as my immediate supervisor, and as such, proved to be an invaluable source of insightful ideas, as well as a formidable foe for the sillier ones.

I have also benefitted greatly from discussions with and the accumulated knowledge of the other members of the Theoretical Quantum Optics Group at Siegen, among them my three office mates Dr. Mazhar Ali, whose passion for knowledge and insight served as a reminder of why I had myself chosen this path in times when going forward got difficult, Marcel Bergmann, whose interests extend far beyond his area of specialization and with whom I fondly remember watching the announcement of the Higgs boson's discovery, and Dr. Costantino Budroni, whose expertise on the topic of quantum contextuality, and whose rare gift to see directly through to the core of an idea, and identify the misconception it rests upon, I have greatly admired. I am furthermore thankful for having had the opportunity of working with Dr. Martin Hofmann, even beyond my time at Siegen, Dr. Tobias Moroder, not just for having the good grace to let me win the occasional game of billards, and Dr. Sabine Wölk.

Additionally, I would be remiss not to express my thanks towards Prof. Dr. Dagmar Bruß, for giving me a new academic home at the Quantum Information Group at the university of Düsseldorf, and who graciously allowed me to take some time to polish this manuscript up for publication. Finally, for comments on this manuscript, I warmly thank Michael Epping.

My trajectory up to this point has not always been a smooth one, and thus, a very special thanks is also due to my family, starting with my father, Bernhard, for never losing faith in me and supporting me through difficult times, no matter what, and to my mother, Barbara, who in many more ways than I am probably aware of provides both the origin of my frame of reference and the initial conditions that serve to define my path in life, and whose too early loss I still deeply mourn. My sisters, Inka and Gisa, I first and foremost wish to thank for, through their lives, giving me an example showing that it can be done after all. You have had much more of an impact on my life than you might know, and I genuinely look up to you.

Last, but by no means least, I want to thank my fiancé, Constance Bartz. Her commitment, love, and support are, more than anything else, what made this work possible—but her bravery in accepting my marriage proposal is what has made me a happy man.

Jochen Szangolies

Contents

List of Illustrations

Figures

Tables

Chapter 1

Introduction and Motivation

NONCONTEXTUALITY may be understood as the idea that properties—of an object, a physical system, or even a theory—should
stand on their own, i.e. be independently fixed; that for every given
property, one can in principle uniquely and definitely determine
whether some object has that property.

Take, for instance, the property of having a certain color: of any
object in our everyday experience, we can definitely say whether or
not it has the property of, say, being blue. Indeed, this ability is to a
large extent what defines the usual notion of objecthood: any object
can in principle be uniquely identified via a list of all its properties.
If that list were sufficiently detailed, the object could be recreated
anytime, anywhere, from nothing but that list, and a supply of the
necessary raw materials. Let us concretely imagine that list as simply
containing the answers to the question 'does the object have property
x?' for all possible properties, which may be represented as simply a
string of 1s and 0s (up to any arbitrary, but finite, degree of accuracy
if we take into account continuously-valued quantities). In this sense,
the list contains all of the object's information.

That in the quantum world, there are no such lists, is one way
of expressing the content of the Kochen-Specker theorem. Basically,
it asserts that, if quantum mechanics—which famously only gives
probabilistic predictions for any observation—were to be completed
with so-called *hidden variables* in order to explain its inherent indeter-
minism as merely a lack of information about the true fundamental
parameters, these hidden variables would have to be contextual: if

each property of a quantum system is to be made definite, then this can only be done in a way such that the assignment of values to every property depends on the context, i.e. the experimental arrangement—for instance, which other properties are investigated simultaneously.

To understand this, one must first recall that, according to quantum mechanics, no list of properties can be complete, since not all can be observed simultaneously up to arbitrary precision. This is essentially the content of Heisenberg's uncertainty theorem. However, the same property—the same observable—may be co-measurable with different sets of observables, i.e. may be observable in different *contexts.*

For a macroscopic analogy, consider a ball drawn from a set of balls which differ along certain characteristics, for instance size: they can be either small or large, mass: they can be either heavy or light, and color: they can be either green or red. Let us stipulate the following correlations between the properties:

- whenever color and size are measured together, the outcome is either that the ball is green and small, or that it is red and large, with equal probability;

- whenever size and mass are measured together, we find a ball that is either small and heavy, or large and light, where both results are again equiprobable; and

- whenever mass and color are measured together, each ball is heavy and red, or light and green, again with the same probability.

Clearly, these assignments are consistent in the sense that whenever a random ball is chosen, it will have a probability of $\frac{1}{2}$ to be green (or red), small (or large), and heavy (or light). However, if we try to observe all three possible properties at the same time, a contradiction arises: the property 'being green' is perfectly correlated with 'being small', which is perfectly correlated with 'being heavy'; however, 'being heavy' is perfectly correlated, in turn, with 'being red'!

This means that there exists no joint probability distribution for *all* the properties of the balls that marginalizes to the distributions for sets of two properties as defined in 1.-3.; even though those distributions marginalize to perfectly consistent single-property probability distributions, such that looking only at those properties by themselves, one would never have noticed anything strange!

The problem discussed above is known as the *marginal problem*, the question of whether or not it is possible to find, for a given set of probability distributions, a common joint distribution that marginializes to these distributions, and our example is a variant of the simplest setting in which its answer is negative [1].

In the setup we have presented, any two properties are simultaneously observable, i.e. *compatible*. However, as we have seen, it is something very different to observe a ball's color in one context with its mass, than it is to observe it in one context with its size: the context of observation matters! It is now the main lesson of the Kochen-Specker theorem that *if* quantum system are to have definite properties at all, then they are like the properties of our hypothetical balls, rather than like the classical macroscopic properties that one can collect in lists.

But does the Kochen-Specker theorem apply to nature? That is, can we actually perform an experiment demonstrating the noncontextual nature of reality—is there actually an appropriate urn from which to draw the balls, so to speak—, and if so, will the quantum mechanical predictions, or classical expectations be supported by this experiment?

These questions and related ones have recently attracted much attention (e.g. [24, 25, 26, 28, 32, 38, 39, 44]), as well as caused a certain amount of controversy [47, 48, 49, 53, 54, 56, 61]. However, great strides have already been made on the experimental front, using photons [41, 34] and neutron interferometry [42, 43] to exhibit quantum violations for specific quantum states, and even in a state-independent way using sequential measurements on trapped ions [45].

Nevertheless, important open questions still remain. In this book, we will be specifically concerned with the so-called *problem of compatibility* [44, 66]: roughly, measurements in the real world are never perfect; but perfect compatibility requires perfect measurements (otherwise, noise effects and accidental couplings to the environment may spoil compatibility), and thus, the notion of contextuality as introduced above does not directly apply. We will attempt to take a step towards resolving this issue.

First, after having provided a brief overview of the theory and history of the subject in chapter 2, in chapter 3 we discuss the behaviour of Kochen-Specker tests under various models of quantum-mechanical noise. This provides bounds on the minimum experimental quality needed in order to theoretically expect to be able to observe the quantum-mechanical violation of noncontextuality inequalities, and thus, the results obtained therein may be used to gauge the likelihood with which an observation of the violation of noncontextuality is genuinely quantum mechanical in nature.

Then, in chapter 4, we attack the problem from the other side, concentrating on finding explicit models of hidden-variable dynamics that are capable of inducing violations of noncontextuality inequalities by randomly changing the hidden-variable state after measurement. The randomness of this change means that it is independent of the measurement context, and in this sense, such models may be considered 'noncontextual'. These models are inspired by considering the actual measurement process: every measurement necessitates an interaction with the measured system; this interaction may introduce arbitrary disturbances on the hidden variable level, over which we have no control. In real experiments, this would show up in the form of measurement errors, i.e. deviations from the ideal quantum mechanical predictions.

Chapter 5 then considers a way out of these troubles by introducing a novel notion of noncontextuality, applicable not to the state, but rather, to the evolution of a system as a whole. This notion is independent of the compatibility of observables, and thus, defined even in the presence of violations of compatibility, i.e. in real experiments. As we explicitly demonstrate, inequalities obeyed by noncontextually evolving systems are violated by quantum mechanics, yet obeyed by the models that were found to violate Kochen-Specker inequalities in chapter 4. Since furthermore, violation of noncontextual evolution implies a violation of Kochen-Specker noncontextuality, because the

Kochen-Specker notion of noncontextuality is a special case of the notion of noncontextual evolution. Thus, experimental verification of such a violation directly implies the validity of the Kochen-Specker theorem.

Finally, in chapter 6, we give a brief summary of the results, consider their implications and point to possible future applications.

Chapter 2

Theory and Background

SINCE its inception in the beginning of the 20th century, quantum mechanics has been subject to continuous discussion and controversy. In this chapter, we will give both a historical and technical overview of some particular aspect of this controversy, namely, the possibility of *completing* quantum mechanics with so-called hidden variables. In particular, we will focus on 'no-go' theorems, which may be used to put empirical limits on any possible completions. We will gradually work our way towards the Kochen-Specker theorem and examine the question of its experimental testability.

2.1 The Completeness of Quantum Theory

In contrast to classical theories, quantum mechanics provides fundamentally probabilistic predictions. Thus, the question of the *completeness* of quantum theory arises: in analogy to classical theories, one might suppose that probabilities only enter into the theory because of our ignorance of the true, fundamental kinematics and/or dynamics. This may be called the *ignorance interpretation* of quantum probability. In order to yield a complete description of reality, quantum mechanics would then have to be supplemented by additional parameters, so-called *hidden variables*.

This question has been raised most famously by Einstein, Podolsky and Rosen (abbreviated EPR) in 1935 [2] (brought into the form most familiar today, referring to spin-entangled electrons, by Bohm and Aharonov in 1957 [3]). EPR define the following *condition of completeness*:

> Every element of the physical reality must have a counterpart in the physical theory. ([2], p. 777)

Thus, their conception of completeness rests on the notion of *elements of reality*. On these, they say the following:

> If, without in any way disturbing a system, we can predict with certainty [...] the value of a physical quantity, then there exists an element of physical reality corresponding to this physical quantity. (*Ibid.*)

Their argument then is simple, yet striking: according to Heisenberg's uncertainty principle, if the observables corresponding to two physical quantities A and B do not commute, i.e. $[A, B] \neq 0$, both quantities cannot simultaneously be measured to arbitrary accuracy. However, they set up an example of two physical systems which, having interacted in the past, must be described by a simultaneous, entangled wave function. They then explain that by measurements on one of the systems, I, the other, II, may be left in an eigenstate of either of two observables, even if they fail to commute. Hence, by their criterion, since system II is not disturbed during the measurement, *both* observables must correspond to an element of physical reality—while naively, the uncertainty principle seems to allow definite reality for at most one of the observables. Thus, they conclude, quantum mechanics must be incomplete[1]. EPR end their discussion with the words:

> While we have thus shown that the wave function does not provide a complete description of the physical reality, we left

---

[1] Actually, they discuss another option: assigning simultaneous reality to two quantities only when both can be simultaneously measured or predicted. However, this would make the reality of a quantity dependent on the measurement, which they discard on the basis that this could not be permitted by any 'reasonable' definition of reality.

open the question of whether or not such a description exists. We believe, however, that such a theory is possible. (*Ibid.*, p. 780)

This further question had, in fact, already been tackled by von Neumann in 1932 [4], in his seminal work on the mathematical foundations of quantum mechanics. In it von Neumann purported to answer this question in general, and in the negative: no completion of quantum mechanics through the introduction of hidden variables is possible. However, in 1966, Bell pointed to a critical shortcoming of the argument [5]. It is instructive to briefly review his version of von Neumann's theorem in order to build a foundation for different 'no-go'-theorems to be discussed later.

Consider two observables A and B of a system, represented in QM by self-adjoint operators (which we will not notationally distinguish from the observables themselves). Then, there exists an observable C such that $C = \alpha A + \beta B$, and if $\langle A \rangle$ and $\langle B \rangle$ denote the expectation values of A and B respectively, then $\langle C \rangle = \alpha \langle A \rangle + \beta \langle B \rangle$ is the expectation value of C. A hidden-variable theory now is committed to the simultaneous existence of definite values $v(A)$, $v(B)$ and $v(C)$ for all three observables (an assumption often referred to as *value definiteness*). Then, one would expect (and von Neumann requires) that $v(C) = \alpha v(A) + \beta v(B)$. But this is generally impossible: let $A = \sigma_x$, $B = \sigma_y$, and $C = \frac{1}{\sqrt{2}}(\sigma_x + \sigma_y)$, with σ_i denoting the familiar Pauli matrices. Then, $v(A)$, $v(B)$ and $v(C)$ may all be either of ± 1. But $\pm 1 \neq \frac{1}{\sqrt{2}}(\pm 1 + \pm 1)$.

However, as Bell explicitly shows, it is possible after all to construct a hidden-variable description of a two-level quantum system. Thus, von Neumann's argument must be in error. In fact, the problem lies with the assumption of the additivity of expectation values for all observables. While this is a property of quantum mechanics, there is no reason to require it of the hidden-variable theory, and Bell's explicit model possesses it only for commuting observables. Bell levels the same criticism at a variant of von Neumann's theorem proposed by Jauch and Piron in [6].

The question of the possibility for a completion of quantum mechanics received its most famous (partial) answer in 1964 by, again, Bell [7]. He proved what today is known simply as *Bell's theorem*, to wit, that if such a more complete description exists, it cannot be local, i.e. dependent only on the events in a system's past lightcone, and agree with quantum mechanics in all instances. To this day, this result forms the paradigm example of a 'no-go' theorem.

Bell's argument proceeds from the Bohm-Aharonov version [3] of the EPR paradox. Consider two two-level quantum systems, for concreteness to be thought of as two spin-$\frac{1}{2}$ particles whose spin $\boldsymbol{\sigma}$ is measured along some direction $\mathbf{n}$. If the system is prepared in the singlet state $|\Psi^-\rangle = \frac{1}{\sqrt{2}}(|\uparrow_{\mathrm{I}} \downarrow_{\mathrm{II}}\rangle - |\downarrow_{\mathrm{I}} \uparrow_{\mathrm{II}}\rangle)$, then, if the spin of particle I is measured along the direction $\mathbf{n}$, measurement of II along the same direction will yield the opposite value, i.e. measurement of $\boldsymbol{\sigma}_{\mathrm{I}} \cdot \mathbf{n}$ yielding 1 implies measurement of $\boldsymbol{\sigma}_{\mathrm{II}} \cdot \mathbf{n}$ yielding -1. This corresponds to the framework of EPR's original argument [2].

Any more complete description, provided by hidden variables collectively denoted $\lambda \in \Lambda$, must then match this behaviour. Take two observers, A and B, each in possession of one of the two particles comprising the EPR pair. Each measures the spin of their particle along some direction, denoted $\mathbf{a}$ and $\mathbf{b}$. Thus, the outcome of each experiment must then be determined by $\mathbf{a}$ and λ, respectively $\mathbf{b}$ and λ, i.e. $A = A(\mathbf{a}, \lambda) \in [-1, 1]$ and $B = B(\mathbf{b}, \lambda) \in [-1, 1]$. If now $p(\lambda)$ is the probability distribution of the hidden variables, we can write the expectation value of their product as

$$\langle AB \rangle \overset{HV}{=} \int_\Lambda d\lambda p(\lambda) A(\mathbf{a}, \lambda) B(\mathbf{b}, \lambda), \qquad (2.1)$$

which must equal the quantum prediction

$$\langle AB \rangle \overset{QM}{=} -\mathbf{a} \cdot \mathbf{b}. \qquad (2.2)$$

From these preliminary considerations, Bell then derives an inequality that all models of this kind (collectively denoted *local realistic*) have to obey. This original 'Bell inequality' is

$$1 + \langle BC \rangle \geq |\langle AB \rangle - \langle AC \rangle| \qquad (2.3)$$

The great importance of Bell's theorem then derives from the fact that utilizing such an inequality, the question of the completion of quantum mechanics by (local) hidden variables becomes accessible to experiment: local realism necessitates a deviation from quantum mechanical predictions in certain situations.

However, Bell's original inequality is not well suited to experiment, since it does not apply in the presence of possible non-detections (i.e. measurements which yield neither $+1$ nor -1). To this end, Clauser, Horne, Shimony and Holt in 1969 proposed an alternative version, known after their initials as the CHSH-inequality [8]:

$$\langle \chi_{\mathrm{CHSH}} \rangle = \langle AB \rangle + \langle BC \rangle + \langle CD \rangle - \langle DA \rangle \leq 2 \qquad (2.4)$$

This inequality, like Bell's original one, holds for all local realistic models. But if the EPR pair is in the state

$$|\Phi^+\rangle = \frac{1}{\sqrt{2}}(|00\rangle + |11\rangle), \qquad (2.5)$$

then, choosing the observables $A = \sigma_x \otimes \mathbb{1}$, $B = -\frac{1}{\sqrt{2}}\mathbb{1} \otimes (\sigma_z + \sigma_x)$, $C = \sigma_z \otimes \mathbb{1}$ and $D = \frac{1}{\sqrt{2}}\mathbb{1} \otimes (\sigma_z - \sigma_x))$ (where $\mathbb{1}$ is the 2×2 unit matrix) yields $\langle \chi_{\mathrm{CHSH}} \rangle = 2\sqrt{2}$ (which value is indeed the maximum attainable for quantum mechanics, known as *Tsirelson's bound* [9]).

With this framework in hand, the first experimental test of a Bell inequality was carried out by Freedman and Clauser in 1972 [10]. Today, the quantum mechanical violation of Bell inequalities is widely accepted, thanks to experiments performed by Aspect and collaborators in 1981-82 [11, 12, 13], and to the 1998 experiment by the group of Zeilinger [14], thus establishing the consensus that local realistic completions of quantum mechanics are indeed ruled out.

2.2 The Kochen-Specker Theorem

It is instructive to inquire into the reason why quantum mechanics violates Bell inequalities. A necessary requirement for Bell-inequality violation is *entanglement*: only states that cannot be written as a tensor product of (pure) subsystem states, i.e. $|\psi_{\text{ent}}\rangle \neq |\psi_1\rangle \otimes |\psi_2\rangle$, may exceed classical bounds. But this is not sufficient: there exist entangled states[2] which nevertheless do not violate any Bell inequality [15, 16]. Thus, non-locality is a property of certain states only. But entanglement is a phenomenon seemingly remote from everyday existence, and therefore one might be tempted to 'shrug off' the implication of Bell's theorem, maintaining that it is of little consequence for most practical purposes. Hence, it would be interesting to investigate whether quantum mechanics as a whole, rather than just some quantum-mechanical states, deviates from classical predictions.

The first step towards just such a result was established by Gleason in 1957 [19]. He proved that on any Hilbert space of dimension greater than three, the only suitable probability measures are given by the density matrices, i.e. that if Π_i is some projector onto a subspace corresponding to the i-th eigenvalue of some observable O, $\mu_\rho(\Pi_i)$, the probability that measurement of O returns i for the state ρ, must be $\text{Tr}(\Pi_i\rho)$, where Tr denotes the trace operation. This is of course nothing else but Born's rule. That in this work lies the germ of an exceptionally strong no-go theorem was first realized by Bell in 1966 [5], who proposed it as a stronger replacement of von Neumann's result consequent on his critique thereof. Earlier, in 1960, Specker had considered similar ideas [20].

As Bell argues, the important feature of Gleason's work with respect to the hidden-variable program is that, since the probability measure provided by density matrices is continuous, *any* assignment of probabilities to properties of some quantum system (represented by projection operators Π_i) must be continuous. However, in a hidden-variable description, only two distinct values, corresponding to the projectors' eigenvalues 0 and 1, which may be interpreted as truth values indicating whether a system possesses a certain property, can occur. Thus, the hidden-variable mapping necessarily contains discon-

---

[2]However, these states have to be mixed—all pure entangled states violate a Bell inequality [17, 18].

tinuities (cf. [21]), and as Bell showed, this entails that two states receiving different values cannot be arbitrarily close together. More explicitly, together with the nonexistence of dispersion-free states[3], Gleason's theorem may be used to demonstrate the nonexistence of a lattice homomorphism between $\mathcal{P}(H)$, the lattice of closed linear subspaces of Hilbert space, and the two-element Boolean algebra $\mathcal{B}_2$ [22]—that is, the nonexistence of a mapping that for every property a quantum system may have uniquely decides whether it does or does not have that property.

Bell then proceeds to subject his theorem to the same sort of criticism he had previously levelled at von Neumann's and Piron and Jauch's argument. His crucial conclusion:

> It was tacitly assumed that measurement of an observable must yield the same value independently of what other measurements may be made simultaneously. ([5], p. 451)

The same spirit is present in [20], where Specker considers 'nonsimultaneously decidable propositions'. This assumption is nowadays generally referred to as *non-contextuality*: the requirement that the question of whether a system has a certain property can objectively be decided without taking into account what other questions are asked (i.e. measurements are performed) simultaneously. Like locality, which it supplants in the present formulation, this seems a sensible requirement, and it is certainly fulfilled for all familiar, macroscopic objects.

The theorem Bell considered in his 1966 paper was given an independent and more definite formulation in 1967 by Kochen and Specker [23]. Their presentation relies on the same crucial insight as Bell's: that rays in Hilbert space having different assignments of the truth values 0 and 1 for some property cannot be arbitrarily close to each other. However, the virtue of their argumentation lies in the explicit construction of a set of rays which, if arranged into a graph

---

[3]A dispersion-free state is a state ρ such that the dispersion $\sigma(O) = \langle O^2 \rangle - \langle O \rangle^2$ vanishes for all operators O.

such that vertexes corresponding to orthogonal rays are joined by an edge, is not true-false colorable, i.e. for which there does not exist a consistent simultaneous assignment of truth values. Basically, while Bell shows that the quantum-mechanical relation $S_x^2 + S_y^2 + S_z^2 = 2 \cdot \mathbb{1}$, where the S_i are the spin observables of a spin-1 particle and $\mathbb{1}$ is the identity operator, cannot always be satisfied using non-contextual hidden variables, Kochen and Specker exhibit an explicit—and most importantly, finite—set of vectors, such that not all of them can fulfill this relation simultaneously.

Before presenting the proof of the theorem, let us first briefly consider its relationship to Bell's 1964 one. Roughly, the Kochen-Specker theorem replaces Bell's assumption of locality with an assumption of non-contextuality. It is easy to show that in certain instances, non-contextuality implies locality [24]: if some observable A can be measured in conjunction with compatible observables $B, C, \ldots$ as well as $L, M, \ldots$, and this can be implemented in such a way that the system may be partinioned into subsystems such that only local manipulations are necessary to implement measurement of either context on either part of the system, then we have the notion of locality as used in Bell's theorem. Furthermore, any Bell inequality can be turned into a Kochen-Specker inequality [25]. Non-contextuality then may be viewed as being more general, and local realistic theories are a subset of non-contextual ones [26]. Also, as will be shown, proofs of the contextuality of quantum theory can be given that do not rely on any specific state being prepared, and thus, are said to be 'state-independent'. In particular, no entanglement is necessary to violate non-contextuality[4].

2.2.1 Kochen and Specker's Original Proof

We will begin by briefly discussing the original proof by Kochen and Specker of their eponymous theorem. This proof, while more involved than more recent examples, is instructive in the sense that it is the

---

[4]In fact, in their original paper [23], Kochen and Specker consider a single-particle realization of their argument.

original example of the 'coloring game' type of proof of the Kochen-Specker theorem. We will here mainly follow the presentation in [27].

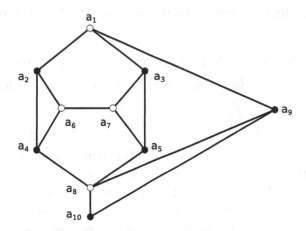

Figure 2.1: $\mathcal{G}_1$: Ten propositions a_i, where simultaneously nonsatisfiable ones are linked by an edge; the coloring shown is inconsistent with the orthogonality constraints, showing that they are incompatible with the requirements $g_1 = 1$ and $a_1 = 1$, but $a_{10} \neq 1$.

Consider first the graph $\mathcal{G}_1$ in Figure 2.1. For the moment, we will consider it as simply having at its vertices certain classical propositions a_i, which are linked by an edge $\{i, j\}$ if a_i and a_j are mutually exclusive, i.e. cannot be both true at the same time. Thus, every edge represents again a proposition:

$$b_{i,j} = \neg(a_i \wedge a_j), \qquad (2.6)$$

where $\neg$ stands for negation, and the wedge $\wedge$ represents the logical **and**. Thus, this proposition is true exactly if at least one of a_i and a_j is false. Similarly, the three triangles in the graph again represent new propositions:

$$c_{ijk} = a_i \vee a_j \vee a_k, \qquad (2.7)$$

where $\vee$ denotes the logical **or**. These propositions are evidently true whenever at least one of a_i, a_j, or a_k is true. Call $\mathcal{E}_1$ the set of all pairs $\{i,j\}$ such that a_i and a_j are linked by an edge, and similarly T_1 the set of all triples $\{i,j,k\}$ such that a_i, a_j and a_k form a triangle in $\mathcal{G}_1$. Now observe that the whole graph represents the following proposition, which is merely the conjunction of all edge and triangle propositions:

$$
\begin{aligned}
g_1 = &b_{1,2} \wedge b_{1,3} \wedge b_{1,9} \wedge b_{2,4} \wedge b_{2,6} \wedge b_{3,5} \wedge b_{3,7} \wedge b_{4,6} \wedge b_{4,8} \wedge b_{5,7} \\
&\wedge b_{5,8} \wedge b_{6,7} \wedge b_{8,9} \wedge b_{8,10} \wedge b_{9,10} \wedge c_{2,4,6} \wedge c_{3,5,7} \wedge c_{8,9,10} \qquad (2.8) \\
\equiv &\bigwedge_{\{i,j\}\in\mathcal{E}_1} b_{ij} \wedge \bigwedge_{\{i,j,k\}\in T_1} c_{ijk}
\end{aligned}
$$

It is now not difficult to see that the truth of g_1, i.e. $g_1 = 1$, together with the truth of a_1, implies the truth of a_{10}: if we assume to the contrary that $g_1 = a_1 = 1$, but $a_{10} = 0$, the truth of $b_{1,2}$, $b_{1,3}$, $b_{1,9}$ and $b_{8,10}$ imply that $a_2 = a_3 = a_9 = 0$, and thus, $a_8 = 1$, since $c_{8,9,10} = 1$. But this implies $a_4 = a_5 = 0$ (because $b_{4,8} = b_{5,8} = 1$), and hence, $a_6 = a_7 = 1$, since $c_{2,4,6} = c_{3,5,7} = 1$ (and we have shown that $a_2 = a_3 = a_4 = a_5 = 0$). But this obviously contradicts $b_{6,7} = 1$; see also the coloring in Figure 2.1.

Consider now the graph $\mathcal{G}_2$ in Figure 2.2, composed of 15 copies of $\mathcal{G}_1$. From $\mathcal{G}_2$, we can construct a proposition g_2 analogous to the way g_1 was constructed from $\mathcal{G}_1$:

$$
g_2 = \bigwedge_{\{i,j\}\in\mathcal{E}_2} b_{ij} \wedge \bigwedge_{\{i,j,k\}\in T_2} c_{ijk}, \qquad (2.9)
$$

where $\mathcal{E}_2$ and T_2 are respectively the edge- and triangle-set of $\mathcal{G}_2$. Using the prior result that $a_1 = 1$ implies $a_{10} = 1$ (and similarly, $a_{18} = 1$ and so on), it is not hard to show that g_2 is always false. Consider the triangle $\{a_1, a_9, a_{41}\}$ in Fig. 2.2. Since $c_{1,9,41} = 1$, at least one of them must be true. Suppose thus $a_1 = 1$. Then, $a_{10} = 1$, $a_{18} = 1$, $a_{26} = 1$, $a_{34} = 1$, and finally, $a_{41} = 1$. However, this contradicts $b_{1,41} = 1$. Thus, since we can perform the same construction starting from a_9 or a_{41}, the proposition g_2 is never satisfiable; alternatively, one says that the graph $\mathcal{G}_2$ is not true/false colorable, i.e. there is no consistent attribution of truth values to the vertices.

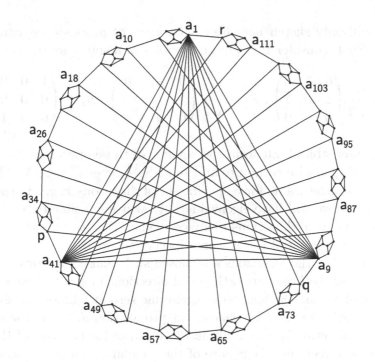

Figure 2.2: $\mathcal{G}_2$: Graph of 117 propositions, where two propositions are again linked by an edge if they cannot be simultaneously satisfied. Note the identifications $p = a_1$, $q = a_{41}$, and $r = a_9$

The crux of the proof is now this: it is possible to find propositions referring to the state of some quantum system such that g_2 is satisfied. Essentially, this means that it is not possible to specify for a quantum system simultaneously all of its properties: we can read the propositions a_i as 'has property a_i', and, as was just discussed, no assignment of truth values to the propositions exists that makes g_2 true.

We will only sketch here how such a set of propositions can be found. First, consider the spin-observables of a spin-1 system:

$$S_x = \frac{1}{\sqrt{2}} \begin{pmatrix} 0 & 1 & 0 \\ 1 & 0 & 1 \\ 0 & 1 & 0 \end{pmatrix}, \; S_y = \frac{1}{\sqrt{2}} \begin{pmatrix} 0 & -i & 0 \\ i & 0 & -i \\ 0 & i & 0 \end{pmatrix}, \; S_z = \begin{pmatrix} 1 & 0 & 0 \\ 0 & 0 & 0 \\ 0 & 0 & 1 \end{pmatrix}$$

(2.10)

The observables built from the squares of these operators, i.e. S_x^2, S_y^2 and S_z^2, pairwise commute and fulfil $S_x^2 + S_y^2 + S_z^2 = 2 \cdot \mathbb{1}$. Thus, they are co-measurable for a specific set of directions x, y, z in space, and exactly two will have eigenvalue 1. Now it is possible to map the propositions a_i to directions in space x_i such that $g_2 = 1$; this then constitues the proof of the Kochen-Specker theorem.

In particular, the directions are chosen such that x_i, x_j correspond to an edge of $\mathcal{G}_2$ if they are orthogonal directions in space; then, since only one of $S_{x_i}^2$ and $S_{x_j}^2$ can have eigenvalue zero, it follows that every b_{ij} necessarily is true if we choose our proposition a_i to be 'the spin in the x_i direction is zero'. Similarly, because for two out of three orthogonal directions, the square of the angular momentum must be one, every c_{ijk} must also be true; but then, g_2 is true, despite being classically false.

The problem of finding such directions can be considered as the problem of 'coloring the sphere': assigning truth values obeying the rules for the $S_{x_i}^2$ to directions as points on the unit sphere. Using the orthogonality relations of $\mathcal{G}_1$, it can be shown that the maximum angle between x_1 and x_{10} equals $\arcsin\left(\frac{1}{3}\right) \approx 19.5°$. This means that between any two directions separated by an angle of not more than $\arcsin\left(\frac{1}{3}\right)$, a graph like $\mathcal{G}_2$ can be constructed, meaning that two differently colored rays cannot be arbitrarily close together. We will briefly sketch the proof.

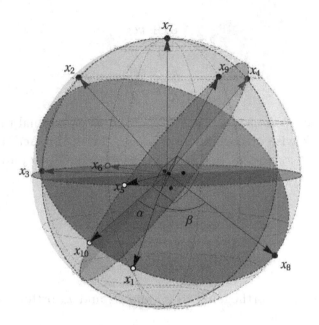

Figure 2.3: Ten spatial directors on the unit sphere realizing the orthogonality relations of $\mathcal{G}_1$, such that the angle α assumes its
maximum value. The orthogonality of x_1 and x_9, x_9 and
x_{10}, x_9 and x_8, as well as x_8 and x_{10} is explicitly indicated.
As a visual aid, the great circles containing $\{x_4, x_5, x_9, x_{10}\}$,
$\{x_2, x_4, x_7\}$, $\{x_3, x_5, x_6\}$, and finally, $\{x_1, x_8, x_{10}\}$ have been
highlighted (shaded disks). The coloring shown is consistent
with the orthogonality relations of $\mathcal{G}_1$, as well as the constraints $g_1 = 1$ and $a_1 = 1$, thus showing how they force
$a_{10} = 1$.

First, call α the angle between x_1 and x_{10}. Then, observe that,
since x_9 is orthogonal to x_1 and x_{10}, and x_8 is orthogonal to x_9 and
x_{10}, this fixes x_8 up to an overall sign; we can thus choose the angle
β between x_8 and x_1 to be equal to $\frac{\pi}{2} - \alpha$, as indicated in Fig. 2.3.
Next, fix an orthonormal triad $\{\hat{e}_1, \hat{e}_2, \hat{e}_3\}$. Identifying x_6 with $\hat{e}_1$ and
x_7 with $\hat{e}_3$, we can write

$$x_2 = \frac{\hat{e}_2 + p\hat{e}_3}{\sqrt{1 + p^2}}$$

$$x_3 = \frac{\hat{e}_1 + q\hat{e}_2}{\sqrt{1 + q^2}}, \tag{2.11}$$

with arbitrary parameters $p, q \in \mathbb{R}$, since x_2 is orthogonal to $x_6 = \hat{e}_1$ and x_3 is likewise orthogonal to $x_7 = \hat{e}_3$. From the fact that x_4 is orthogonal to x_6 and x_2, and that x_5 is orthogonal to x_3 and x_7, we get that

$$x_4 = \frac{x_6 \times x_2}{|x_6 \times x_2|} = \frac{\hat{e}_3 - p\hat{e}_2}{\sqrt{1 + p^2}}$$

$$x_5 = \frac{x_3 \times x_7}{|x_3 \times x_7|} = \frac{\hat{e}_2 - q\hat{e}_1}{\sqrt{1 + q^2}}. \tag{2.12}$$

Similarly, with x_1 orthogonal to x_2 and x_3 and x_8 orthogonal to x_5 and x_4, we obtain

$$x_1 = \frac{x_2 \times x_3}{|x_2 \times x_3|} = \frac{-\hat{e}_3 + p\hat{e}_2 - pq\hat{e}_1}{\sqrt{1 + p^2 + p^2q^2}}$$

$$x_8 = \frac{x_5 \times x_4}{|x_5 \times x_4|} = \frac{-\hat{e}_1 - p\hat{e}_2 - pq\hat{e}_3}{\sqrt{1 + q^2 + p^2q^2}}. \tag{2.13}$$

We are now able to calculate the angle between x_1 and x_8, and hence, between x_1 and x_{10}. With

$$\cos(\beta) = x_1 \cdot x_8 = \frac{pq}{\sqrt{(1 + p^2 + p^2q^2) \cdot (1 + q^2 + p^2q^2)}} \tag{2.14}$$

we obtain

$$\sin(\alpha) = \cos(\frac{\pi}{2} - \alpha) = \frac{pq}{\sqrt{(1 + p^2 + p^2q^2) \cdot (1 + q^2 + p^2q^2)}}, \tag{2.15}$$

which assumes its maximum value at $p = q = 1$, where $\sin(\alpha) = \frac{1}{3}$. Thus, for any $\alpha \leq \arcsin\frac{1}{3}$, a construction realizing the orthogonality

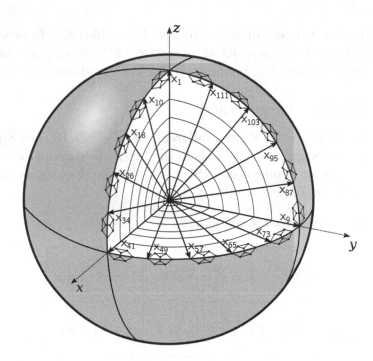

Figure 2.4: The 15 spatial directions used in the proof of the Kochen-Specker theorem.

relations of the graph $\mathcal{G}_1$ is possible between x_1 and x_{10}, showing that both must receive the same color, or equivalently, that the propositions a_1 and a_{10} must receive the same truth value.

Let us now consider the sphere octant shown in Figure 2.4. With the construction as shown, we can prove that a_1, a_9 and a_{41} must be colored the same, as was already discussed above: five copies of $\mathcal{G}_2$ divide each right angle into parts of $18°$. If we thus start at a_1, we find that a_{10} must receive the same color, as must a_{18}, and so on. However, since x_1, x_9 and x_{41} are orthogonal spatial directions, out of $S_{x_1}^2$, $S_{x_9}^2$ and $S_{x_{41}}^2$, two must be colored true, while one must be colored false. This then establishes the contradiction: there exists no consistent true/false coloring for the 117 directions x_i, and thus, no consistent assignment of values to the observables $S_{x_i}^2$. Hence, for a spin-1 quantum system ρ, not all properties of the form 'ρ has angular momentum squared 1 in the direction x_i' can simultaneously be fixed.

2.2.2 The Peres-Mermin Square

Another, conceptually slightly different, proof of the Kochen-Specker theorem can be given on the basis of the Peres-Mermin square [29, 30]. This is an array of nine observables on a four-level quantum system, arranged as in Table 2.1.

Table 2.1: The Peres-Mermin square.

$A = \sigma_x \otimes \mathbb{1}$	$B = \mathbb{1} \otimes \sigma_x$	$C = \sigma_x \otimes \sigma_x$
$a = \mathbb{1} \otimes \sigma_y$	$b = \sigma_y \otimes \mathbb{1}$	$c = \sigma_y \otimes \sigma_y$
$\alpha = \sigma_x \otimes \sigma_y$	$\beta = \sigma_y \otimes \sigma_x$	$\gamma = \sigma_z \otimes \sigma_z$

As one readily verifies, the observables in each row and column commute, and the product of the observables in each row equals $\mathbb{1}$ (here, the 4×4 identity matrix), as do the products of the first two colums. However, in the last column, $Cc\gamma = -\mathbb{1}$. Thus, it is not consistently possible to assign the values ± 1 to the observables, as a non-contextual hidden variable theory would demand: the row products necessitate an even number of -1s, while the column products require an odd number, thus producing a contradiction similar to that in the previous section. It is important to note that this proof is wholly independent of the state of the quantum system: as we have only ever talked about observables, the conclusion must hold for any quantum state, even, for instance, the maximally mixed one $\rho = \frac{\mathbb{1}}{\text{Tr}(\mathbb{1})}$.

An important, but somewhat subtle requirement in this proof is that of *compatibility*: in order to be able to make meaningful assertions about their simultaneous values, the observables in each row and column must be co-measurable, i.e. it must be possible to obtain perfect information about all their values simultaneously. Without this requirement, it would be nonsensical to talk about the value of the product of observables, since their values would not all be definite at once.

2.3 Testing the Kochen-Specker Theorem

As we have already discussed, one of the main virtues of Bell's theorem is that it makes the question of the completion of quantum mechanics by hidden variables accessible to experiment[5]. It would certainly be desirable to claim the same success for the Kochen-Specker theorem; however, as will be discussed, its experimental testing faces even greater challenges and controversies.

Let us first consider Kochen and Specker's original proposal to implement their scheme using measurements of a spin-1 particle. They consider measuring the squared spin components S_x^2, S_y^2 and S_z^2 of an orthohelium atom placed in an electric field of appropriate (rhombic) symmetry. In this context, there exists a single observable, the perturbation Hamiltonian H_s, measurement of which reveals the values of S_x^2, S_y^2 and S_z^2. Since $S_x^2 + S_y^2 + S_z^2 = 2 \cdot \mathbb{1}$, two of these values must be 1, while one is 0.

However, this does not work as a direct test of non-contextuality: only one orthogonal triplet, i.e. one context, is considered at any given time, and thus, we cannot say anything about the value of an observable in distinct contexts. The same reservation applies to the direct testing of the observables of the Peres-Mermin square: the observable A cannot be tested in the contexts ABC and $Aa\alpha$

---

[5]As reported in [31], nobel laureate E. M. Purcell, in a lecture delivered at Harvard University, "expressed his delight at having lived long enough to see a philosophical problem settled in the laboratory".

simultaneously, since non-compatible observables cannot be measured on the same physical system; a disagreement between these tests might then reveal nothing more than a difference in the measurement procedure.

To resolve this difficulty, Cabello and García-Alcaine [32] proposed a scheme in which non-contextual theories make predictions contrary to quantum mechanics for every single system, independently of its quantum state. In their original formulation, the system was considered to be two spin-$\frac{1}{2}$ particles; a version considering spin- and path-degrees of freedom of a single particle was proposed by Simon et al. [33], and experimentally realized by Huang et al. in 2002 [34]. The results were in strong agreement with quantum mechanics; nevertheless, for reasons to be discussed in sections 2.5 and 2.6, there is to date no unanimous agreement on the decisiveness of this and similar tests.

2.4 Non-Contextuality Inequalities

A different route to the testability of the Kochen-Specker theorem is provided by deriving inequalities, conceptually similar to those used in Bell tests. Early work on the subject of testing the KS-theorem using inequalities was performed by Roy and Singh in 1993 [35], who introduced the notion of 'stochastic' non-contextuality in order to apply an inequality of the CHSH form (2.4); in a similar vein, Basu et al. in [36] consider applying the CHSH inequality to the spin- and path-degrees of freedom of a single spin-$\frac{1}{2}$ particle.

The first inequalities specifically applicable to the Kochen-Specker theorem were suggested by Simon, Brukner and Zeilinger [37] and Larsson [38], who independently considered Kochen and Specker's original proposal, extending it for the case of imprecisely specified measurements (s. a. sect. 2.5). Non- contextuality inequalities in full analogy to Bell inequalities were proposed by Cabello et al. [26], as well as by Klyachko, Can, Binicioğlu, and Shumovsky [39]. Finally, the first state-independent non-contextuality inequalities were derived by Cabello in 2008 [40].

An important novelty in the inequality-based approach to the Kochen-Specker theorem is the realization that quantum theory gives predictions different from those of noncontextual hidden-variable theories even for *sequential* measurements, as long as they are compatible [44]; thus, an expression such as $\langle ABC \rangle$ can be regarded as simply an instruction to measure the observables in order on a system, and form the product of the observed values.

In the following, we will be mainly concerned with two inequalities: one is the already familiar CHSH inequality, interpreted as a non-contextuality inequality, and the other is a state-independent inequality first proposed in [40].

2.4.1 The CHSH-Inequality

It is easy to see that the CHSH inequality 2.4 is applicable to non-contextual as well as local realistic theories [46]. Consider the CHSH-operator:

$$\chi_{\text{CHSH}} = AB + BC + CD - DA \qquad (2.16)$$

The CHSH inequality $\langle \chi_{\text{CHSH}} \rangle \leq 2$ can be violated only if $\|\chi_{\text{CHSH}}\| > 2$, where $\| \cdot \|$ denotes the operator norm. Since χ_{CHSH} is a self-adjoint operator, $\|\chi_{\text{CHSH}}\|^2 = \|\chi^2_{\text{CHSH}}\|$. Thus, a sufficient condition for violating the inequality is $\|\chi^2_{\text{CHSH}}\| > 4$. The square of the CHSH-operator evaluates to

$$\chi^2_{\text{CHSH}} = 4 + (AC - CA)(BD - DB) = 4 + [A, C][B, D], \quad (2.17)$$

where we have used the fact that since they are ± 1-valued, the observables square to 1. Clearly, now, if all observables are considered to be simple random variables, and thus, non-contextual, they necessarily commute. Hence, the assumption of non-contextuality implies the CHSH-inequality.

2.4.2 An Inequality from the Peres-Mermin Square

A drawback of the inequality discussed in the previous section is its state-dependence: only for certain quantum states is χ_{CHSH} in fact greater than two in quantum mechanics. However, the proof of the Kochen-Specker theorem presented in subsection 2.2.2 yields a testable inequality almost directly. Recall that in the Peres-Mermin square (2.1), it was impossible to simultaneously satisfy the constraints imposed by the products of the observables along the rows as well as along the columns. This impossibility may be considered to arise from the final column, where the product of all measurements necessarily yields -1. Thus, we can simply collect the rows and columns into a single expression [40]:

$$\langle \chi_{PM} \rangle = \langle ABC \rangle + \langle abc \rangle + \langle \alpha\beta\gamma \rangle + \langle Aa\alpha \rangle + \langle Bb\beta \rangle - \langle Cc\gamma \rangle \quad (2.18)$$

For any non-contextual theory, this value is bounded by four: $\langle \chi_{PM} \rangle \leq 4$. However, quantum mechanics predicts $\langle \chi_{PM} \rangle = 6$, irrespective of the quantum state.

2.5 The Finite-Precision Problem

The inequalities discussed in the previous section were, at least in part, proposed to answer a specific criticism levelled at the possibility of experimentally testing the Kochen-Specker theorem. This criticism is known as the *finite precision problem*: basically, since every real measurement is specified only up to a certain finite precision, it is impossible to be absolutely certain that the observable one set out to measure is in fact the observable that is being measured. But if such an uncertainty exists, it is always possible to find a subset of directions such that its observables are colourable in the Kochen-Specker sense [47, 48, 49].

2.5.1 MKC Models

An early argument proposing that any Kochen-Specker set of un-colourable vectors may be arbitrarily closely approximated by a co-lourable set, which thus is not distinguishable from the former by measurement, is due to Pitowsky [50, 51]. However, his analysis relied on the axiom of choice and the continuum hypothesis, and may be considered unconvincing on these grounds. Later on, building on work by Godsil and Zaks [52], Meyer [47] proposed that a 2-colourable set of rational unit vectors may be used for such an approximation in the setting orginally considered by Kochen and Specker; this result was generalized to arbitrary Hilbert spaces by Kent [48]. Subsequently, Clifton and Kent constructed an explicit model to reproduce the quantum-mechanical predictions using non-contextual hidden variables [49]. Such models have since come to be known MKC-models after their originators.

2.5.2 Answers to the Finite-Precision Problem

The MKC models have provoked a substantial amount of interesting discussion, of which only a brief outline can be given here. Roughly, the proposed rejoinders may be grouped into three different strategies: (1) denying the physical plausibility of the MKC sets, (2) rejecting MKC's conclusion, and (3) extending the Kochen-Specker theorem to encompass also scenarios with imprecisely specified measurements.

The first strategy is followed by Cabello [53], who argues that the theorem applies to measured values rather than possessed ones, and points out that a model taking the rational unit sphere as its physical state space has the curious property that, for instance, not all theoretically allowed superpositions of states are physically possible. Similarly, Havlicek et al. [54] consider the non-closedness of the rational sphere to be problematic: certain operations, such as for instance the logical nor, may then in certain cases yield results not physically allowed.

Following the second line of argument, Appleby [55] has argued that the MKC-models themselves are, actually, contextual, but in a peculiar way: not the value of a given observable depends on the context, but its very existence. Also, Mermin [56] has argued that the continuity of probability (see the discussion of Gleason's theorem in sect. 2.2) spoils the argumentation of Meyer, Kent, and Clifton: effects due to the finite precision of measurement simply wash out over large enough measurement numbers. On the other hand, Cabello [57] has argued that MKC's colourable sets in fact lead to predictions that differ observably from those of quantum mechanics. Finally, Cabello and Larsson [58] have constructed an explicit example of a set of rational vectors that violate the inequality derived in [39].

The third route of investigation seems to be the most promising one. Largely, it consists of attempts to find a version of the Kochen-Specker theorem that applies to imprecisely specified observables, possibly in a stochastic sense. One such attempt is made by Breuer [59], who uses POVMs (positive operator-valued measures [60]) in order to define finite-precision observables; unfortunately, his proposal does not apply to the MKC-models directly, since it depends on a rotational symmetry that those models do not possess. Following a different approach, Larsson [38] and Simon et al. [37], as has already been mentioned, derive inequalities whose violation indicates a violation of non-contextuality. The attractiveness of their scheme is that it is framed in operational terms: basically, their framework amounts to a setting in which the experimenter has access to a black box, which has three knobs, corresponding to the observables S_x^2, S_y^2 and S_z^2, that can be set to various orientations. Based on the setting of these knobs, the box produces a certain result, in the form of three numbers, each assigned to one of the knobs. If this result can be understood in such a way that the outcome for a given knob always only depends on its position, and is therefore independent of the positions of the other knobs, the theory is non-contextual; otherwise, non-contextuality is violated. Note that no reference to either quantum mechanics or the precision of measurement was made in this definition. A similar approach is taken by Basu et al. [36], who derive the CHSH inequality

2.4 from the assumption of non-contextuality alone[6]. In figure 2.5, an operational CHSH-setup is depicted.

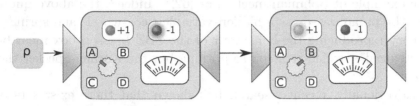

Figure 2.5: An operational setup for testing noncontextuality via the CHSH-inequality: the system, ρ, is fed through the two measurement apparata. If the outcome of either box is independent of the setting of the other, its behaviour can be explained by noncontextual hidden variables; contrariwise, the impossibility to do so establishes contextuality without reference to Kochen-Specker colorability.

Against this, in [61], Barrett and Kent have levelled the objection that

> there is nothing specifically non-classical about a black box that is behaving SBZ-contextually. One could easily construct such a box out of cog-wheels and springs. Thus with no knowledge of or assumptions about the internal workings of the box, one could not use it to distinguish classical from quantum behaviour.

This is certainly true. However, the same objection can be raised against any quantum experiment, since it is always possible to simulate local quantum effects classically, if perhaps only at the cost of exponential inefficiency. Furthermore, this does not change the fact that the behaviour of the black box is contextual; whether this contextuality is implemented quantum-mechanically or by some sophisticated, but essentially classical machinery is a question of interpretation

---

[6]Similarly, the derivation presented above also does not assume any part of the quantum formalism.

(and that one can always find an interpretation that is classical in the sense of assigning definite values to observables is proven by the example of Bohmian mechanics [62]). Indeed, the above quote may be paraphrased as "a hidden-variable ('cog-wheels and springs') theory reproducing the behaviour of the SBZ-Larsson box must be contextual"—which is of course nothing but the Kochen-Specker theorem.

Additionally, recent research has shown that there exist sets of vectors that are not Kochen-Specker sets—i.e. that are true-false colourable—, and which nevertheless can be used to derive inequalities that are obeyed by noncontextual theories, yet violated by quantum mechanics [63, 64, 65].

2.6 The Problem of Compatibility

According to the discussion of the previous section, inequalities such as 2.4 appear to be the most promising route to a definite test of the Kochen-Specker theorem. However, there is another problem which seems to block the way towards such a test that has to be discussed. This problem is known as the *problem of compatibility*, and it has its roots in the fact that contextuality is only defined for perfectly compatible, and thus, co-measurable, observables [44, 66], as has already been stressed.

For present purposes, the notion of compatibility is best defined in operational terms: call a set of observables $\{A, B, C, \ldots\}$ *compatible* if, in any sequence of measurements of observables from this set, the value of every observable remains constant; alternatively, the value of an observable, say A, is not disturbed if any of the other observables are measured. More explicitly, two observables A and B can be called compatible if for any measurement sequence $\mathcal{S}_{AB} \in \{A, B, AA, AB, BA, BB, AAA, AAB, \ldots\}$ the values of A and B agree, no matter in which position within the sequence the observable is measured, i.e. $v(A_i|\mathcal{S}_{AB}) = v(A_j|\mathcal{S}_{AB})$ and $v(B_i|\mathcal{S}_{AB}) = v(B_j|\mathcal{S}_{AB})$

for all positions i, j and sequences $\mathcal{S}_{AB}$. Compatibility for multiple observables is defined in a fully analogous way.

It is now plain to see where the problem of compatibility lies: in any real experiment, noise introduced either via interactions with the environment, interactions of separate qubits with each other, or imperfectly implemented unitary transformations, will typically cause a violation of compatibility to some small, but non-zero degree, i.e. it can in general not be guaranteed that in some sequence of measurements $ABAC \ldots A$ the first and last measurement of A will agree, even though all measured observables are in principle compatible.

2.6.1 A Kochen-Specker Test on Separated Qutrits

One reply to the problem of compatibility was proposed by Cabello and Terra Cunha [66]. They propose to utilize a system of spatially separated qutrits, on which measurement of a Kochen-Specker inequality is performed. Measurements within one context are carried out on different qubits, in order to ensure their compatibility. However, one could imagine several objections that might be raised against this scheme. First, even if we assume that the measurements are perfectly isolated and hence cannot possibly influence one another, interactions with the environment still might lead to violations of compatibility, in the sense that measurements of AB and BA do not necessarily agree. Furthermore, both measurements can, in principle, influence one another even if both systems are spatially separated, if the influence is mediated non-locally; so even though the authors argue that their inequalities should not be viewed as Bell inequalities, one could maintain that only local realistic theories are excluded by their proposed experiment—there may in principle exist a non-local noncontextual theory that accounts for all measurement outcomes.

2.6.2 Extended KS Inequalities

Another approach was taken by Gühne et al. [44]. They propose extended Kochen-Specker inequalities, in which additional 'error' terms are introduced to compensate for possible incompatibilities. First, they show that, for any two observables AB, $\langle AB \rangle \leq \langle A_1 B_2 \rangle + 2 p^{\text{flip}}[AB]$, where $p^{\text{flip}}[AB]$ denotes the probability that the measurement of A disturbs the observable B, i.e. flips it from a predetermined value to its opposite. Thus,

$$p^{\text{flip}}[AB] = p[(B_1^+|B_1) \text{ and } (B_2^-|A_1 B_2)] + p[(B_1^-|B_1) \text{ and } (B_2^+|A_1 B_2)]. \tag{2.19}$$

Here, the numerical indices refer to the position of the measurement of some observable within a sequence of such measurements, which sequence is denoted as the condition, while the signs indicate the outcome of the measurement. So, for instance, $p[(B_2^-|A_1 B_2 B_3)]$ indicates the probability that the outcome of the measurement of the observable B is -1, given that this measurement was the second in the sequence ABB (n. b. that in a setting where compatibility is violated, the probability for obtaining -1 for the third measurement may well be different!). This makes it possible to obtain an extended CHSH inequality that is valid even in the presence of compatibility violations:

$$\langle \chi_{\text{CHSH}} \rangle \leq 2(1 + p^{\text{flip}}[AB] + p^{\text{flip}}[BC] + p^{\text{flip}}[CD] + p^{\text{flip}}[DA]) \tag{2.20}$$

However, the probabilities $p[(B_1^+|B_1) \text{ and } (B_2^-|A_1 B_2)]$ are not experimentally accessible, since one can measure B either first, or second, but not both. In order to include only measurable quantities, they then derive an upper bound to the flip-probabilities; to do so, they make the following assumption:

Assumption 2.1. *(Cumulative noise.) Additional measurements only increase the amount of disturbance suffered by the system. Thus:*

$$
\begin{aligned}
p[(B_1^+|B_1) \text{ and } (B_2^-|A_1 B_2)] &\leq\ p[(B_1^+|B_1) \text{ and } (B_1^+, B_3^-|B_1 A_2 B_3)] \\
&\equiv\ p[(B_1^+, B_3^-|B_1 A_2 B_3)] \tag{2.21}
\end{aligned}
$$

The reasoning behind this assumption is the following: if measuring one observable, A_1, disturbs the state such that measuring B_2 produces a different outcome than measuring B_1 would have, then it stands to reason that more measurements only increase the disturbance, such that measuring B_3 in the sequence $B_1A_2B_3$ has an even greater probability from differing from B_1. This assumption will be further examined, and counterexamples considered, in chapter 4.

This term now is accessible to experiment: one can simply measure the sequence $B_1A_2B_3$ enough times to obtain an estimate for the probability that B_3 differs from B_1. If one then defines error terms of the form $p^{\text{err}}[B_1A_2B_3] = p[(B_1^+, B_3^-|B_1A_2B_3)]+p[(B_1^-, B_3^+|B_1A_2B_3)]$, a measurable extended CHSH inequality can be formulated:

$$\langle\chi_{\text{CHSH}}\rangle \leq 2(1 + p^{\text{err}}[B_1A_2B_3] + p^{\text{err}}[C_1B_2C_3]$$
$$+ p^{\text{err}}[D_1C_2D_3] + p^{\text{err}}[A_1D_2A_3])$$
$$(2.22)$$

If the above assumption 2.1 holds, then the violation of this inequality implies a violation of noncontextuality even if the observables are not perfectly compatible, i.e. under realistic experimental conditions.

Chapter 3

Noise-Robustness of Kochen-Specker Tests

In order to understand the details of experimental tests of the Kochen-Specker theorem, we will first perform an analysis of the noise-robustness of certain tests that have been performed. To do so, we will break down the measurement process into a series of discrete steps, and allow the system to perform a noisy evolution, according to certain well-known models for experiment-induced noise (see sects. 3.3-3.6). This will allow us to provide bounds on the minimum quality needed for an experiment in order to conclusively establish a violation of non-contextuality.

3.1 The Measurement Process

Let us first consider how the introduction of noise into the measurement process may lead to violations of compatibility. Recall that we had defined compatibility in an operational way as the repeatability of individual measurements within measurement sequences, see section 2.6.

However, experimental imperfections imply that the above ideal situation can never be achieved in practice. Ambiguities in state preparation/detection, imperfectly implemented unitary transformations, and interactions with the environment, to name a few examples, generally spoil perfect compatibility. One may consider this to be due to noise influences acting on the state between measurements. This situation is schematically illustrated in Figure 3.1.

3.2 Noise Models and Quantum Operations

As it is depicted in Figure 3.1, the disturbance of a system by measurement- (or, more generally, environment-)induced noise may be modelled by sending the system through a noisy channel, effecting the transformation $\rho \to \mathcal{E}(\rho)$, if the system was originally in the state ρ [60].

In order to develop this model, consider first the dynamics of a closed quantum system, i.e. some arbitrary state ρ_S evolving unitarily:

$$\rho_S \to \rho_S' = U \rho_S U^\dagger. \tag{3.1}$$

If, now, the system is not closed, but is part of a larger system together with some environment ρ_E, then in general the evolution of the total system, restricted to the system of interest by tracing out the environmental part, will no longer be unitary. For the combined system, the evolution then is

$$\rho_S \otimes \rho_E \to U(\rho_S \otimes \rho_E)U^\dagger; \tag{3.2}$$

the evolution of the system under consideration on its own is then given by

$$\mathcal{E}(\rho_S) : \rho_S \to \rho_S' = \mathrm{tr}_E \left[U(\rho_S \otimes \rho_E)U^\dagger \right], \tag{3.3}$$

which defines the *quantum operation* $\mathcal{E}$, and where tr_E denotes the partial trace with respect to the environment ρ_E. Quantum operations, especially those used to model noisy evolution of a quantum system, are also sometimes referred to as *(noisy) quantum channels*, because of their formal similarity to classical noisy information channels [60].

This representation, while intuitive, is somewhat inconvenient to work with mathematically. Thus, it is useful to introduce the so-called *operator-sum representation*, using the quantum channel's *Kraus operators* [67]. For this, we first assume that the environment can be

Perfectly compatible observables:

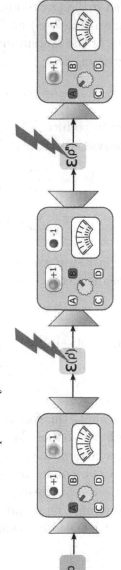

Noise-induced compatibility violation:

Figure 3.1: Noise-induced violations of compatibility.

considered to be in a pure state, $\rho_E = |e_0\rangle\langle e_0|$. This we can always do, since even if the environment is actually in a mixed state, we can purify using a (ficticious) additional system, which does not change the dynamics of the system under consideration [60]. Thus, equation 3.3 can be written as

$$\rho'_S = \mathrm{tr}_E \left[U(\rho_S \otimes |e_0\rangle\langle e_0|)U^\dagger \right]. \tag{3.4}$$

If we now introduce a basis $\{|e_i\rangle\}$ for the environment, we can compute the partial trace, yielding

$$\begin{aligned} \rho'_S &= \sum_i \langle e_i| \left[U(\rho_S \otimes |e_0\rangle\langle e_0|)U^\dagger \right] |e_i\rangle \\ &\equiv \sum_i E_i \rho_S E_i^\dagger, \end{aligned} \tag{3.5}$$

where in the last step we have introduced the operator-sum representation by means of the Kraus operators $\{E_i = \langle e_i|U|e_0\rangle\}$. From the condition

$$\mathrm{Tr}\left(\mathcal{E}(\rho)\right) = 1, \tag{3.6}$$

we immediately obtain the relation

$$\sum_i E_i E_i^\dagger = \mathbb{1}. \tag{3.7}$$

In order to apply this formalism to the problem at hand, we need to develop it a little further. We are concerned mainly with expectation values of sequences of measurements of the form $\langle ABC\ldots\rangle$, with the property that subsequent measurements are performed on the system after it has been sent through a noisy channel. The expectation value of one measurement on a state after it has been sent through a noisy channel $\mathcal{E}$ can be readily evaluated:

$$\langle A\rangle = \mathrm{Tr}(A\mathcal{E}(\rho)) \tag{3.8}$$

However, in case of a sequence $\langle AB\rangle$, where A is measured on the original state, whilst measurement of B takes place on the state after it has been subject to noise effects, a little more work is needed. First, we must find the state after the first measurement. If we have obtained, say, the outcome +1, the state after measurement is

$$\rho_{A-} = \frac{\Pi_A^- \rho \Pi_A^-}{\mathrm{Tr}(\Pi_A^- \rho)}, \tag{3.9}$$

where Π_A^- denotes the projector onto the eigenspace of A to the eigenvalue -1. The probability of finding, say, $B = +1$ after having found $A = -1$ then is

$$p(B = +1 | A = -1) = \mathrm{Tr}\left(\Pi_B^+ \frac{\Pi_A^- \rho \Pi_A^-}{\mathrm{Tr}(\Pi_A^- \rho)}\right). \tag{3.10}$$

This post-measurement state is then sent through the noisy channel $\mathcal{E}$. The expectation value can be written:

$$\langle AB \rangle = p_{A+B+} - p_{A+B-} - p_{A-B+} + p_{A-B-}, \tag{3.11}$$

where for instance $p_{A+B+} = p(B = +1, A = +1)$ is shorthand for 'the probability of obtaining the outcomes $A = +1$, $B = +1$', etc. With eq. 3.10, we get then

$$\begin{aligned}
\langle AB \rangle &= p(A^+)\mathrm{Tr}\left(\Pi_B^+ \mathcal{E}\left\{\frac{\Pi_A^+ \rho \Pi_A^+}{\mathrm{Tr}(\Pi_A^+ \rho)}\right\}\right) \\
&\quad - p(A^+)\mathrm{Tr}\left(\Pi_B^- \mathcal{E}\left\{\frac{\Pi_A^+ \rho \Pi_A^+}{\mathrm{Tr}(\Pi_A^+ \rho)}\right\}\right) \\
&\quad - p(A^-)\mathrm{Tr}\left(\Pi_B^+ \mathcal{E}\left\{\frac{\Pi_A^- \rho \Pi_A^-}{\mathrm{Tr}(\Pi_A^- \rho)}\right\}\right) \\
&\quad + p(A^-)\mathrm{Tr}\left(\Pi_B^+ \mathcal{E}\left\{\frac{\Pi_A^- \rho \Pi_A^-}{\mathrm{Tr}(\Pi_A^- \rho)}\right\}\right)
\end{aligned} \tag{3.12}$$

Now, we can use that, for instance, $p(A^-) = \mathrm{Tr}(\Pi_A^- \rho)$, which because of the linearity of $\mathcal{E}$ and the trace cancels with the normalization factor:

$$\begin{aligned}
\langle AB \rangle &= \mathrm{Tr}\left(\Pi_B^+ \mathcal{E}\left\{\Pi_A^+ \rho \Pi_A^+\right\}\right) - \mathrm{Tr}\left(\Pi_B^- \mathcal{E}\left\{\Pi_A^+ \rho \Pi_A^+\right\}\right) \\
&\quad - \mathrm{Tr}\left(\Pi_B^+ \mathcal{E}\left\{\Pi_A^- \rho \Pi_A^-\right\}\right) + \mathrm{Tr}\left(\Pi_B^- \mathcal{E}\left\{\Pi_A^- \rho \Pi_A^-\right\}\right)
\end{aligned} \tag{3.13}$$

The expectation value of a longer measurement sequence then is analogously

$$\langle ABC \rangle = Tr \left[C\mathcal{E} \left\{ \Pi_B^+ \mathcal{E}(\Pi_A^+ \rho \Pi_A^+ - \Pi_A^- \rho \Pi_A^-)\Pi_B^+ \right. \right.$$
$$\left. \left. - \Pi_B^- \mathcal{E}(\Pi_A^+ \rho \Pi_A^+ - \Pi_A^- \rho \Pi_A^-)\Pi_B^- \right\} \right] \tag{3.14}$$

Thus, we now have the machinery to compute the expectation values of arbitrary measurement sequences subject to different kinds of noises.

However, we will also want to compute the effect of noisy measurements on the error terms introduced in section 2.6.2. In order to do so, we must first analyze their form. An error term such as

$$p^{\mathrm{err}}[B_1 A_2 B_3] = p[(B_1^+, B_3^- | B_1 A_2 B_3)] + p[(B_1^-, B_3^+ | B_1 A_2 B_3)] \tag{3.15}$$

quantifies the probability of the observable B flipping its value due to a measurement of A. Evidently, this is the sum of the probability for obtaining the outcome -1 for the second measurement of B, after having obtained $+1$ as the result of the first measurement, and the probability for obtaining $+1$ for the second measurement of B, where the first measurement yielded -1. Let us thus focus on just the term $p[(B_1^-, B_3^+ | B_1 A_2 B_3)]$, which we abbreviate as p_{B-AB+}.

Evidently, $p_{B-AB+} = p_{B-A+B+} + p_{B-A+B+}$. To calculate now, say, p_{B-A+B+}, first recall that the probability of observing $B = -1$ in the state ρ is

$$p_{B-} = \mathrm{Tr}(\Pi_B^- \rho), \tag{3.16}$$

which measurement outcome leaves the system in the state $\rho_{B-} = \frac{\Pi_B^- \rho \Pi_B^-}{\mathrm{Tr}(\Pi_B^- \rho)}$. Thus, the probability of observing first $B = -1$, and then $A = +1$ is

$$p_{B-A+} = p_{A+|B-} p_{B-} = \mathrm{Tr}(\Pi_A^+ \mathcal{E}(\rho_{B-}))\mathrm{Tr}(\Pi_B^- \rho), \tag{3.17}$$

analogously to 3.10, 3.12. After this measurement, the system is in the state $\rho_{B-A+} = \frac{\Pi_A^+ \mathcal{E}(\rho_{B-})\Pi_A^+}{\mathrm{Tr}(\Pi_A^+ \mathcal{E}(\rho_{B-}))}$ Consequently, the probability of observing the sequence $B = -1$, $A = +1$, $B = +1$, works out to:

$$p_{B-A+B+} = p_{B+|A+B-}p_{B-A+}$$
$$= \mathrm{Tr}(\Pi_B^+ \mathcal{E}(\rho_{B-A+}))\mathrm{Tr}(\Pi_A^+ \mathcal{E}(\rho_{B-}))\mathrm{Tr}(\Pi_B^- \rho), \qquad (3.18)$$

and analogously for p_{B-A-B+}.

3.3 Depolarizing Noise

A special, very general type of quantum noise is the *depolarizing channel* [60]. Essentially, it corresponds to a process by which the quantum system is, with a certain probability p, replaced by the completely mixed state $\frac{1}{\mathrm{Tr}(1)}$, while it is left invariant with probability $1 - p$. Thus, the system's evolution is

$$\mathcal{E}(\rho) = p\frac{1}{\mathrm{Tr}(1)} + (1 - p)\rho \qquad (3.19)$$

If we restrict our attention to a single qubit as object system, the channel can be written as

$$\mathcal{E}(\rho) = \frac{p}{4}(\sigma_x \rho \sigma_x + \sigma_y \rho \sigma_y + \sigma_z \rho \sigma_z) + (1 - \frac{3p}{4})\rho, \qquad (3.20)$$

where the σ_i are the Pauli matrices. From this, we can directly read off the Kraus operators:

$$E_0 = \sqrt{1 - \frac{3p}{4}}1$$

$$E_1 = \sqrt{\frac{p}{4}}\sigma_x$$

$$E_2 = \sqrt{\frac{p}{4}}\sigma_y \qquad (3.21)$$

$$E_3 = \sqrt{\frac{p}{4}}\sigma_z$$

This channel acts on a two-qubit system as follows:

$$\mathcal{E}_{\text{dep}}(\rho) = (\mathcal{E}^1_{\text{dep}} \otimes \mathcal{E}^1_{\text{dep}})(\rho) = \sum_{i,j=0}^{3} (E_i \otimes E_j)\rho(E_i \otimes E_j)^\dagger, \qquad (3.22)$$

where $\mathcal{E}^1_{\text{dep}}$ denotes the single-qubit depolarizing channel. Thus, the Kraus operators of the two-qubit depolarizing channel are simply

$$E_{ij} = E_i \otimes E_j, \qquad (3.23)$$

In this form, we can now apply the depolarizing channel to several Kochen-Specker -inequalities and investigate their behaviour under noisy measurements.

3.3.1 The CHSH-Inequality

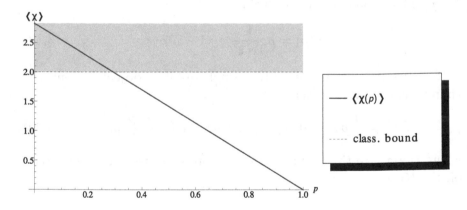

Figure 3.2: The CHSH inequality subject to depolarizing noise. The shaded region shows where an exclusion of noncontextuality is possible.

Under the effect of depolarizing noise, the value of $\langle \chi_{\text{CHSH}} \rangle$ experiences a linear correction:

$$\langle \chi_{\text{CHSH}} \rangle^{\text{Dep}} = 2\sqrt{2} - 2\sqrt{2}p \qquad (3.24)$$

Thus, as is shown in Figure 3.2, at a certain point the noise effects will make the detection of a quantum violation of the CHSH inequality impossible. We can interpret this as a restriction on the minimum quality of the experiment needed to detect such a violation; in this case, the depolarisation probability must fulfil the condition

$$p_{\text{Dep}} < \left(1 - \frac{1}{\sqrt{2}}\right) \approx 0.293. \tag{3.25}$$

3.3.2 The Extended CHSH-Inequality

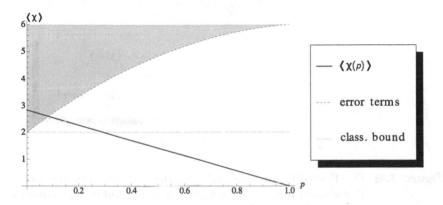

Figure 3.3: The CHSH inequality subject to depolarizing noise, together with the error terms. Again, the shaded region shows where an exclusion of noncontextuality is possible.

In extending the above analysis to the CHSH inequality extended with error terms, 2.22, as was noted above, we have to take into account the dependency of these error terms on the amount of added noise. This yields much more stringent constraints on the required experimental quality. In the present case, the extended CHSH inequality takes the form

$$\langle \chi_{\text{CHSH}} \rangle^{\text{Dep}} \leq 2 + \left(8 - \frac{1}{\sqrt{2}}\right)p + \left(\frac{1}{\sqrt{2}} - 4\right)p^2 \tag{3.26}$$

meaning that the depolarization probability must fulfil

$$p_{\text{Dep}} < \frac{16 + 3\sqrt{2} - \sqrt{434 - 48\sqrt{2}}}{2\sqrt{2} - 16} \approx 0.084 \qquad (3.27)$$

in order to still detect a violation. This is shown in Figure 3.3.

3.3.3 The Peres-Mermin Inequality

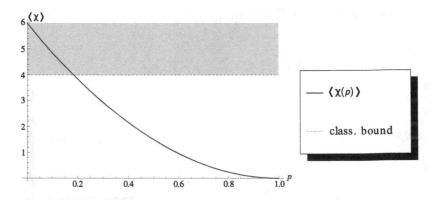

Figure 3.4: The Peres-Mermin inequality subject to depolarizing noise. The shaded region shows where an exclusion of noncontextuality is possible.

The same methods may be applied to studying the noise-robustness of the Peres-Mermin inequality (2.18). Due to the double application of the depolarizing channel, the correction here is quadratic in the probability:

$$\langle \chi_{\text{PM}} \rangle^{\text{Dep}} = 6(p - 1)^2 \qquad (3.28)$$

This is shown in Figure 3.4. To detect violations of noncontextuality, the depolarization probability must obey

$$p_{\text{Dep}} < \frac{1}{3}(3 - \sqrt{6}) \approx 0.184. \qquad (3.29)$$

3.4 Bit-Flipping

A simple type of error that may be introduced during the evolution of a quantum state is the *bit flip* [60]. As the name implies, this corresponds simply to the flipping of a state to an orthogonal one with a certain probability p. The action of this channel on the pure states $|1\rangle\langle1|$ and $|0\rangle\langle0|$ is therefore:

$$\begin{aligned}
\mathcal{E}_{BF}(|1\rangle\langle1|) &= (1-p)|1\rangle\langle1| + p|0\rangle\langle0| \\
\mathcal{E}_{BF}(|0\rangle\langle0|) &= (1-p)|0\rangle\langle0| + p|1\rangle\langle1|
\end{aligned} \tag{3.30}$$

This can be achieved using the following Kraus operators:

$$\begin{aligned}
E_0 &= \sqrt{1-p}\begin{pmatrix} 1 & 0 \\ 0 & 1 \end{pmatrix} = \sqrt{1-p}\mathbb{1} \\
E_1 &= \sqrt{p}\begin{pmatrix} 0 & 1 \\ 1 & 0 \end{pmatrix} = \sqrt{p}\sigma_x
\end{aligned} \tag{3.31}$$

3.4.1 The CHSH-Inequality

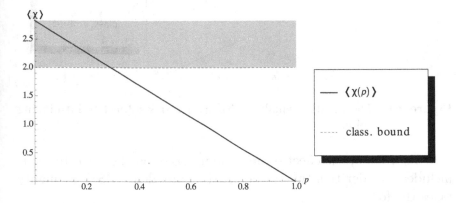

Figure 3.5: The CHSH inequality subject to bit-flipping noise.

Applied to the CHSH-inequality, the bit-flip channel produces a correction of the form

$$\langle\chi_{\mathrm{CHSH}}\rangle^{\mathrm{BF}} = 2\sqrt{2} - 2\sqrt{2}p. \tag{3.32}$$

Thus, as is also shown in Figure 3.5, no violation of noncontextuality can be observed unless the bit-flip probability obeys

$$p_{\mathrm{BF}} < 1 - \frac{1}{\sqrt{2}} \approx 0.293. \tag{3.33}$$

Remarkably, for the CHSH inequality, bit-flip errors thus induce the same kind of behaviour as depolarizing noise does. However, differences exist with respect to the extended CHSH and Peres-Mermin inequalities, as will be discussed in the next two sections.

3.4.2 The Extended CHSH-Inequality

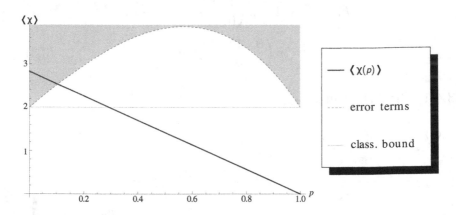

Figure 3.6: The CHSH inequality with error terms subject to bit-flipping noise.

In Figure 3.6, the effect of bit-flipping noise on the error terms is included. Under this type of noise, the extended CHSH-inequality takes the form

$$\langle \chi_{\mathrm{CHSH}} \rangle^{\mathrm{BF}} \leq 2 - 2(\sqrt{2} - 4)p + (5\sqrt{2} - 8)p^2 - 3\sqrt{2}p^3. \tag{3.34}$$

This yields a bound on the bit-flipping probability of

$$p_{\mathrm{BF}} \lesssim 0.105. \tag{3.35}$$

3.4.3 The Peres-Mermin Inequality

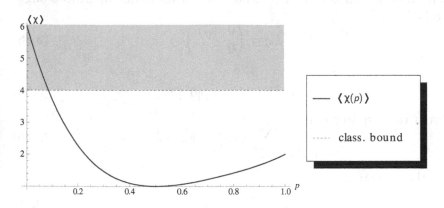

Figure 3.7: The Peres-Mermin inequality subject to bit-flipping noise.

The correction suffered by the Peres-Mermin inequality under the influence of noise of the bit-flipping type is the following:

$$\langle \chi_{\text{PM}} \rangle^{\text{BF}} = 6 - 28p + 56p^2 - 48p^3 + 16p^4 \tag{3.36}$$

Thus, in order to observe a violation of noncontextuality, we need for

$$p_{\text{BF}} < 0.085 \tag{3.37}$$

to hold (no simple closed form seems to exist). This is shown in Figure 3.7.

3.5 Amplitude Damping

The *amplitude damping* channel is a type of noise that characterizes the effect of energy dissipation on a system. This models processes such as the spontaneous emission of a photon, or the attenuation of light in an optical cavity [60].

This channel takes an excited state, $|1\rangle\langle 1|$, to a deexcited one, $|0\rangle\langle 0|$, with a certain probability p. Thus, one of its Kraus operators ought to be

$$E_0 = \begin{pmatrix} 0 & \sqrt{p} \\ 0 & 0 \end{pmatrix}, \qquad (3.38)$$

since

$$E_0|1\rangle\langle 1|E_0^\dagger = p|0\rangle\langle 0|. \qquad (3.39)$$

From the requirement

$$\sum_i E_i E_i^\dagger = \mathbb{1}, \qquad (3.40)$$

we then get that

$$E_1 = \begin{pmatrix} 1 & 0 \\ 0 & \sqrt{1-p} \end{pmatrix}. \qquad (3.41)$$

Thus, we see that the application of this channel to the state $|1\rangle\langle 1|$ results in

$$\mathcal{E}_{AD}(|1\rangle\langle 1|) = E_0|1\rangle\langle 1|E_0^\dagger + E_1|1\rangle\langle 1|E_1^\dagger = p|0\rangle\langle 0| + (1-p)|1\rangle\langle 1|, \qquad (3.42)$$

while applied to the state $|0\rangle\langle 0|$, we simply get

$$\mathcal{E}_{AD}(|0\rangle\langle 0|) = |0\rangle\langle 0|, \qquad (3.43)$$

i.e. an excited state is deexcited with probability p, while a non-excited state is left invariant.

Again, applied to a system of two qubits, the action of the channel is:

$$\mathcal{E}_{AD}(\rho) = (\mathcal{E}_{AD}^1 \otimes \mathcal{E}_{AD}^1)(\rho) = \sum_{j,i=0}^{1} (E_i \otimes E_j)\rho(E_i \otimes E_j)^\dagger, \qquad (3.44)$$

yielding the Kraus operators

$$E_{00} = E_0 \otimes E_0 = \begin{pmatrix} 0 & 0 & 0 & p \\ 0 & 0 & 0 & 0 \\ 0 & 0 & 0 & 0 \\ 0 & 0 & 0 & 0 \end{pmatrix}$$

$$E_{01} = E_0 \otimes E_1 = \begin{pmatrix} 0 & 0 & \sqrt{p} & 0 \\ 0 & 0 & 0 & \sqrt{p}\sqrt{1-p} \\ 0 & 0 & 0 & 0 \\ 0 & 0 & 0 & 0 \end{pmatrix}$$

$$E_{10} = E_1 \otimes E_0 = \begin{pmatrix} 0 & \sqrt{p} & 0 & 0 \\ 0 & 0 & 0 & 0 \\ 0 & 0 & 0 & \sqrt{p}\sqrt{1-p} \\ 0 & 0 & 0 & 0 \end{pmatrix} \quad (3.45)$$

$$E_{11} = E_1 \otimes E_1 = \begin{pmatrix} 1 & 0 & 0 & 0 \\ 0 & \sqrt{1-p} & 0 & 0 \\ 0 & 0 & \sqrt{1-p} & 0 \\ 0 & 0 & 0 & 1-p \end{pmatrix}$$

3.5.1 The CHSH-Inequality

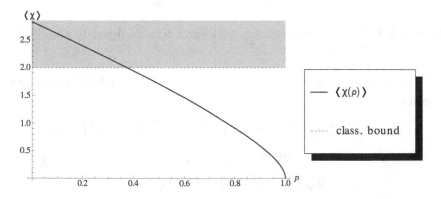

Figure 3.8: The CHSH inequality subject to amplitude-damping noise.

Under amplitude damping noise, the CHSH inequality receives the correction

$$\langle \chi_{\text{CHSH}} \rangle^{\text{AD}} = \sqrt{2}(1 - p + \sqrt{1-p}). \tag{3.46}$$

Accordingly, as shown in Figure 3.8, the amplitude damping probability (i.e. the probability of energy losses to the environment) must obey

$$p_{AD} < \frac{1}{2}\left(1 - 2\sqrt{2} + \sqrt{1 + 4\sqrt{2}}\right) \approx 0.376. \tag{3.47}$$

3.5.2 The Extended CHSH-Inequality

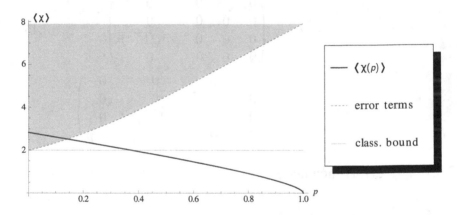

Figure 3.9: The CHSH inequality with error terms subject to amplitude-damping noise.

Taking the error terms into account, the extended CHSH inequality becomes

$$\langle \chi_{\text{CHSH}} \rangle^{\text{AD}} \leq \frac{1}{8}(16 + (44 - 11\sqrt{2} - \sqrt{2 - 2p} - 4\sqrt{1-p})p$$
$$- (4 - 30\sqrt{2} - \sqrt{2 - 2p} \tag{3.48}$$
$$- 2\sqrt{1-p})p^2 - (4 + 11\sqrt{2})p^3);$$

thus, in order to observe violations of noncontextuality, we need

$$p_{\text{AD}} \lesssim 0.143. \tag{3.49}$$

3.5.3 The Peres-Mermin Inequality

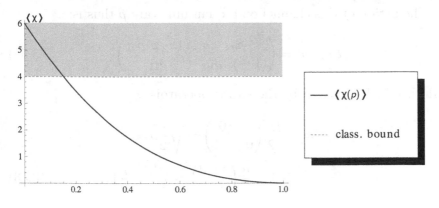

Figure 3.10: The Peres-Mermin inequality subject to amplitude-damping noise.

For the Peres-Mermin inequality, amplitude damping induces the correction

$$\langle \chi_{\mathrm{PM}} \rangle^{\mathrm{AD}} = (1-p)(2+4\sqrt{1-p}-(4+3\sqrt{1-p})p-6p^2+2p^3). \quad (3.50)$$

Thus, the amplitude damping probability must obey

$$p_{\mathrm{AD}} < 0.151 \quad (3.51)$$

in order for a violation of noncontextuality to be experimentally observable, as depicted in Figure 3.10.

3.6 Phase Damping

The quantum channel known as *phase damping* models a noise effect that is uniquely quantum-mechanical in nature: the loss of phase information incurred via, for instance, random scattering of a photon within a waveguide [60]. Thus, this channel provides a simple model for decoherence: the off-diagonal entries in the density matrix decay away, indicating the 'leaking' of quantum information into the environment, while the diagonal entries are left invariant.

The action of this channel on a quantum state ρ thus is:

$$\mathcal{E}_{PD}(\rho) = \begin{pmatrix} \rho_{00} & (1-p)\rho_{01} \\ (1-p)\rho_{10} & \rho_{11} \end{pmatrix}, \tag{3.52}$$

which may be realized by the Kraus operators

$$\begin{aligned} E_0 &= \sqrt{\frac{p}{2}} \begin{pmatrix} 1 & 0 \\ 0 & -1 \end{pmatrix} = \sqrt{\frac{p}{2}}\sigma_z \\ E_1 &= \sqrt{1-\frac{p}{2}} \begin{pmatrix} 1 & 0 \\ 0 & 1 \end{pmatrix} = \sqrt{1-\frac{p}{2}}\mathbb{1}. \end{aligned} \tag{3.53}$$

3.6.1 The CHSH-Inequality

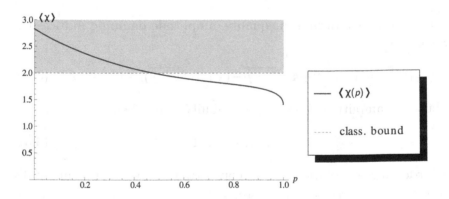

Figure 3.11: The CHSH inequality subject to decoherence.

Under phase damping noise or decoherence, the CHSH inequality receives a correction of the form

$$\begin{aligned} \langle \chi_{CHSH} \rangle^{PhD} = \frac{1}{\sqrt{2}} \Big(&p^4 - 2p^3 + \left(\sqrt{1-p}+3\right)p^2 \\ &- \left(\sqrt{1-p}+2\right)p + 2\left(\sqrt{1-p}+1\right) \Big), \end{aligned} \tag{3.54}$$

and thus, the decoherence probability must obey

$$p_{\text{PhD}} < 0.478. \tag{3.55}$$

This is depicted in Figure 3.11.

3.6.2 The Extended CHSH-Inequality

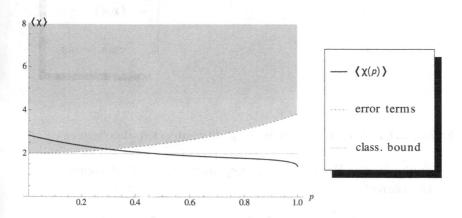

Figure 3.12: The CHSH inequality with error terms subject to phase-damping noise.

In the case of decoherence, the extended CHSH inequality becomes:

$$
\begin{aligned}
\langle \chi_{\text{CHSH}} \rangle^{\text{PhD}} \leq \frac{1}{32} (&(10 - 3\sqrt{2})p^8 + 4(3\sqrt{2} - 10)p^7 \\
&+ (\sqrt{2 - 2p} - 30\sqrt{2} + 108)p^6 \\
&+ (-3\sqrt{2 - 2p} + 42\sqrt{2} - 164)p^5 \\
&+ (7\sqrt{2 - 2p} - 41\sqrt{2} + 202)p^4 \\
&+ (-7\sqrt{2 - 2p} + 22\sqrt{2} - 148)p^3 \\
&+ 6(\sqrt{2 - 2p} - \sqrt{2} + 16)p^2 + 64),
\end{aligned}
\tag{3.56}
$$

meaning that in order to observe a violation of noncontextuality, we need for

$$p_{\text{PhD}} \lesssim 0.306, \tag{3.57}$$

as can also be gleaned from Figure 3.12.

3.6.3 The Peres-Mermin Inequality

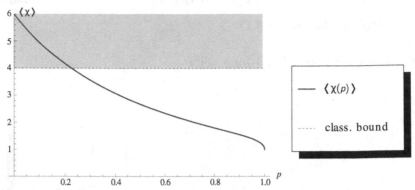

Figure 3.13: The Peres-Mermin inequality subject to decoherence.

Finally, for the Peres-Mermin inequality subject to decoherence, we get the correction

$$
\begin{aligned}
\langle \chi_{\mathrm{PM}} \rangle^{\mathrm{PhD}} = \frac{1}{2}(&4 + 8\sqrt{1-p} - 4(2+3\sqrt{1-p})p \\
&+ (15 + 16\sqrt{1-p})p^2 - (21 + 16\sqrt{1-p})p^3 \\
&+ 2(11 + 6\sqrt{1-p})p^4 - (17 + 6\sqrt{1-p})p^5 \\
&+ 2(5 + \sqrt{1-p})p^6 - 4p^7 + p^8
\end{aligned}
\tag{3.58}
$$

This amounts to the following condition on the decoherence probability:

$$
p_{\mathrm{PhD}} < 0.226
\tag{3.59}
$$

This is shown in Figure 3.13.

3.7 Summary

As we have seen, under a wide variety of noise models, the quantum violation of Kochen-Specker inequalities is strictly decreasing. Thus, certain requirements on experimental quality must be met in order to observe a violation of noncontextuality; however, in no case do these requirements seem unmeetably strong.

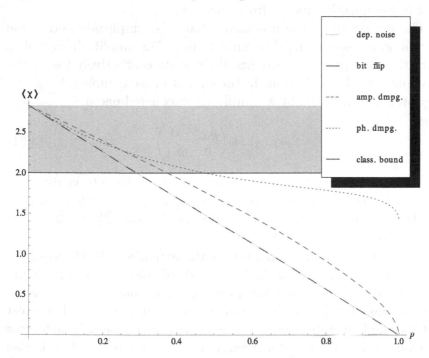

Figure 3.14: The CHSH inequality under different noise models.

In Figure 3.14, the dependence of the violation of the CHSH-inequality on a noise-parameter p for certain kinds of noises is shown. For two kinds of noises, depolarization and bit-flipping errors, the dependence is simply linear, acting to destroy all correlations. This is readily appreciable: for depolarizing noise, in the case of total depolarization, the state is simply replaced by the completely mixed one $\rho = \frac{1}{\text{Tr}(\mathbb{1})}$, on which the measurements yield random results. For the bit-flip channel, it is easy to show that the state $\rho = |\Phi^+\rangle\langle\Phi^+|$ is invariant under the action

$$\rho \rightarrow (\sigma_x \otimes \sigma_x)\rho(\sigma_x \otimes \sigma_x). \tag{3.60}$$

But for $p = 1$, this is just the action of the bit-flipping channel (all other Kraus operators vanish). But then, the second measurement is just independently random from the first.

The behaviour of the nonlinear channels, amplitude and phase damping, is also not hard to understand. For amplitude damping, once the noise parameter reaches 1, the state is effectively lost to the environment. The behaviour in the case of phase damping, for $p = 1$, any state ρ is replaced by a completely decohered one, i.e.

$$\mathcal{E}_{PhD}(\rho) = \begin{pmatrix} \rho_{00} & 0 \\ 0 & \rho_{11} \end{pmatrix}. \tag{3.61}$$

But this does not imply that the correlations have to vanish; and indeed, in the present case, it is easy to verify that, while $\langle AB \rangle = \langle DA \rangle = 0$, $\langle BC \rangle = \langle CD \rangle = \frac{1}{\sqrt{2}}$, and thus, $\langle \chi_{\text{CHSH}} \rangle_{p=1}^{\text{PhD}} = \sqrt{2}$, as can also be seen in Figure 3.14.

The dependence of the bound for the extended CHSH inequality on the noise parameter for different kinds of noise is shown in Figure 3.15. It is interesting that for maximum noise parameter, the error terms vanish for the bit-flip channel, since, again, for $p = 1$, we just get the original state back. For the depolarizing channel, each error probability is equal to $\frac{1}{2}$ for maximum noise parameter. This is also not surprising: after all, for maximum depolarization, we are left with the completely mixed state in the end, and thus, the final measurement is simply random, and has thus an even chance of agreeing with the first.

Lastly, in Figure 3.16, we see the behaviour of the Peres-Mermin inequality subject to the different kinds of noises discussed in this chapter. Broadly, the dependence on the noise parameter is qualitatively similar for all noise models as in the case of the CHSH-inequality, though the induced constraints are considerably more severe. We have collected the constraints for all inequalities and all kinds of noises studied in Table 3.1.

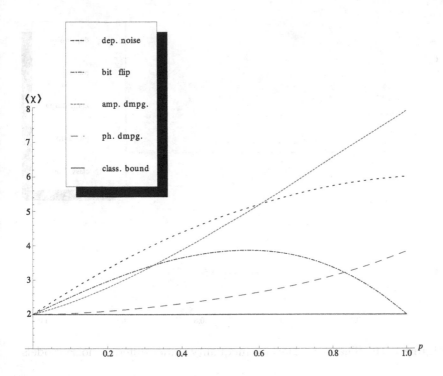

Figure 3.15: The extended CHSH-inequality under different noise models.

Table 3.1: Maximum values for the noise parameter in order to still be able to detect a violation of noncontextuality.

	CHSH	CHSH ext.	Peres-Mermin
Depolarization	0.293	0.084	0.184
Bit-Flip	0.293	0.105	0.085
Ampl. Damping	0.376	0.143	0.158
Phase Damping	0.478	0.306	0.226

As can clearly be seen in the direct comparison, the constraints placed on experimental quality provided by the error-term extended inequalities are the most strict ones for depolarizazion and amplitude damping, while for the bit-flip and phase damping channels, the noise

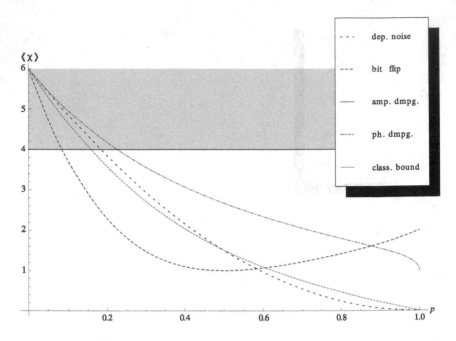

Figure 3.16: The Peres-Mermin inequality under different noise models.

influence is strongest for the Peres-Mermin inequality. In the case of the bit-flip channel, it is interesting to note that for the Peres-Mermin inequality, the constraint is nearly twice as strong as for any other kind of noise.

Chapter 4

Statistical Models for Noisy Measurements

IN the previous chapter, we have seen how quantum noise may influence the experimental tests of Kochen-Specker inequalities. The aim of this chapter is to consider the measurement process from the perspective of an explicit model acting on the space of possible hidden variables, in order to see how this affects measurement results. In particular, we will model the influence of noise on the measurement process using a purely stochastic—and hence, explicitly non-contextual—process, acting on the states of the hidden variables.

4.1 Conceptual Model of the Measurement Process

In this chapter, we will aim to provide a suitable model of the influence of noise on the hidden-variable state, and thus, the measurement results. In order to maintain the non-contextuality of the model, it is necessary that it does not depend on the choices of previous measurements, i.e. that it has no 'memory'. This will be achieved by simply considering our hidden-variable state to be a probability distribution over possible measurement outcomes, which is updated in a random way after each measurement. This simple model will turn out to be surprisingly powerful: as will be shown, despite being arguably non-contextual, it can lead to violations of Kochen-Specker inequalities that can be seen to be due to the incompatibilities induced by the model, which parallel those present in every real experiment.

4.1.1 Ontic State Space

The first notion we will need is that of an *ontic state*. An ontic state may be considered to be a state of the hidden variables that uniquely determines the value of measurable observables, i.e. that characterises the properties of the system in an unambiguous way [68]. Due to this uniqueness, it is possible to characterise the state by simply listing all its properties, or at least all those of interest in some given experiment.

Since we are exclusively interested in cases in which the properties of interest are given by the values of dichotomic observables, we can thus represent any ontic state by an ordered list of these values, symbolized by $+$ and $-$, i.e. any given state $\lambda \in \Lambda$ is of the form $(++-+--\cdots)$, denoting the values of observables $A, B, C, D, \ldots$ in sequence. In order to streamline notation, it is useful to interpret each such list as a binary number, and use its decimal value as an index referring to the state, such that, for instance, the state $(+ + - +)$ may be uniquely referred to as λ_2.

This essentially provides us with a partition of the ontic state space: any given λ_i refers to the set (or equivalence class) of hidden variable states yielding the same outcome i for a set of observables. Evidently, different hidden variables described by the same λ_i cannot be experimentally distinguished. This takes heed of the fact that the observables we measure may be, in fact, coarse-grained rather than truly microscopic properties of the system.

Another way to view these ontic states which will be useful later is offered by considering the following basis:

$$(+) = \begin{pmatrix} 1 \\ 0 \end{pmatrix}, \quad (-) = \begin{pmatrix} 0 \\ 1 \end{pmatrix} \tag{4.1}$$

Then, any given state λ_i can be considered the tensor product of these basis states, e.g.:

$$\lambda_9 = (-+-+-) = (-) \otimes (+) \otimes (+) \otimes (-) \tag{4.2}$$

4.1.2 Violations of Compatibility

Given the possibility of a hidden-variable description underlying quantum theory, we must take into account our ignorance of the hidden-variable dynamics; in particular, it is possible that interactions during measurement may lead to uncontrollable, and therefore essentially random, changes in the state of the hidden variables. For instance, during measurement, the act of measuring may disturb a system in the state λ_i, causing it to transition to some other state λ_j. This, of course, entails a possible violation of compatibility, as an observable previously measured as $+$ may, upon repeated measurement within some sequence of measurements, now yield $-$ (cf. the definition of compatibility in section 2.6). This is schematically represented in Figure 4.1 for a system in which we are interested in two observables, A and B, and can thus partition the hidden-variable space into the four sets $\lambda_0 = (++)$, $\lambda_1 = (+-)$, $\lambda_2 = (-+)$, and $\lambda_3 = (--)$, where the values of A and B are indicated in sequence.

The evolution of the hidden-variable state depicted in Figure 4.1 is clearly deterministic; however, due to our ignorance of the detailed dynamics and kinematics of the hidden variables, it is clear that we must consider more general cases. For instance, two states leading to the same measurement results for all observables may nevertheless evolve differently during a sequence of measurements; thus, we must allow for probabilistic splitting of the evolution. Also, we do not necessarily have complete control over the hidden-variable state we prepare. Therefore, we will in general only be able to initialize the system with a certain probability distribution $p(\lambda)$ over the hidden variables. An example of such a more general evolution is given in figure 4.2.

Let us use the concrete evolution in figure 4.2 to calculate the expectation value $\langle ABA \rangle$ of the measurement sequence. To do this, we need to assign probabilities to all possible evolutions of the system.

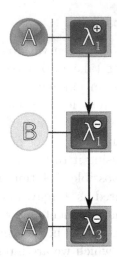

Figure 4.1: Schematic representation of a sequence of measurements,
together with the evolution of the hidden variables. The physical system is indicated by the underlying grey square, while
the λ_i designate its state. The result of the measurement of
the observable to the left on the system in some given state is
indicated by the value in the circle. The transition from λ_1 to
λ_3 causes the second measurement of A to disagree with the
first, and thus, violates compatibility. Time runs downward
throughout.

There are five of these:

$$\lambda_1 \to \lambda_1 \to \lambda_1$$
$$\lambda_1 \to \lambda_0 \to \lambda_3$$
$$\lambda_0 \to \lambda_1 \to \lambda_1$$
$$\lambda_0 \to \lambda_0 \to \lambda_3$$
$$\lambda_0 \to \lambda_2 \to \lambda_1$$

For simplicity, let us assume that all these are equally probable,
i.e. $p_{111} = p_{103} = p_{011} = p_{003} = p_{021} = \frac{1}{5}$. The value of ABA for

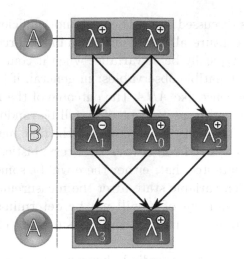

Figure 4.2: A more general evolution: starting from a state that may be either λ_1 or λ_0, all of the depicted transitions are possible. Thus, during repeated runs of the same sequence of measurements, the values of the observables need not agree, and their expectation values will generally differ from ± 1.

each evolution is simply the product of the value of each observable in the respective state, i.e. $A_1 B_0 A_3 = (+1)(+1)(-1) = -1$, where the notation A_i means 'the value of A in the state λ_i. Thus, the expected value is

$$\langle ABA \rangle = \sum_{ijk} p_{ijk} A_i B_j A_k = -\frac{3}{5}. \qquad (4.3)$$

It may seem odd that we have elected to assign probabilities to entire evolutions *in toto*, rather than simply giving probabilities to each branch, or, more accurately, specifying a conditional distribution over the λ_i at each measurement. The reason is that this would impose an unwarranted restriction on the hidden-variable dynamics, effectively assuming that the distribution over possible evolutions is always factorizable, i.e. that (in the present case) $\langle ABA \rangle = \langle A \rangle \langle B \rangle \langle A \rangle$.

The model as discussed so far is clearly not sufficient, in the sense that it cannot capture all quantum-mechanical predictions. Most seriously, it will typically be at variance when it comes to the case of explicitly non-compatible observables: in general, if $[A, B] \neq 0$, for a measurement sequence like ABA, the outcome of the B-measurement as well as the second measurement of A will be random according to quantum mechanics, while the outcome for the final measurement of A in a sequence like $ABAA$ will always be deterministic. But the model clearly cannot guarantee that: either there will be some probability to change the hidden variable state after the measurement of A—then, the second measurement of A will not be deterministic. Or, there is no such possibility—but then, the measurement of B will not be random.

This problem can be remedied, however, by noting that in the case of explicit incompatibility between two observables, there is no good reason to assume noncontextuality (cf. Bell's objection to von Neumann's no-go theorem in sect. 2.1). Thus, since there always exists a possible evolution for the system compatible with the quantum predictions, we can simply introduce an additional selection rule which picks out such an evolution in the case of measurements of explicitly noncompatible observables. This selection rule will be strongly contextual, and in fact, depend on the entire sequence of measurements that is performed; but this is not at variance with the implications of the Kochen-Specker theorem, which requires noncontextuality only in the case of compatible observables. Thus, we can use a two-tiered dynamics to bring the model into account with experimental observations: in the case of compatible measurements, the model simulates the effects of experimental imperfections; the probabilities for state changes can always be chosen such that they are in agreement with observation. If one measures incompatible observables instead, then an evolution is chosen that is in complete agreement with the expected results.

Furthermore, note that we have introduced, in sect. 2.5.2, an operational framework for contextuality tests in order to evade the finite-precision problem. In this sense, we should imagine that we are provided with a box, which produces certain outputs from certain input settings. The question of contextuality is then simply the question of whether the behaviour of the box can be explained in terms of a noncontextual theory or not—i.e. whether its outputs depend on its inputs individually, or can only be explained by considering all inputs together. In this framework, since it is in principle independent of the details of quantum mechanics, the notion of incompatible observables does not strictly apply, and the model discussed above is adequate to represent a certain class of noncontextual theories that might be put forward to explain the box's behaviour.

4.2 Markov Models

In this section, we aim to present a more explicit model that corresponds to a specific sub-class of the models discussed above. In particular, we will demonstrate that, using this model, a violation of both the usual CHSH inequality (2.4) and the extended one (2.22) is possible, despite its noncontextuality—recall that the hidden-variable state transition is effected independently of the context.

The model will be implemented by considering the hidden-variable states λ_i as the states of a *Markov chain*, that is, as the states in a process such that a memoryless, probabilistic transition is implemented between them. Here, the memorylessness guarantees noncontextuality: transitions depend only on the state the system is currently in. These transitions are effected by *stochastic matrices* [69]. A (left-) stochastic matrix is an $n \times n$ matrix $M = (m_{ij})$ such that

$$\sum_i^n m_{ij} = 1. \tag{4.4}$$

This condition ensures that for each probability vector $\mathbf{p} = (p_i)$, for which $\sum_i^n p_i = 1$, $\mathbf{p}' = M\mathbf{p} = (\sum_j^n m_{ij}p_j)$ again is a probability vector such that $\sum_i p_i' = 1$. It is the classical equivalent of the normalization condition 3.6 for quantum channels.

In this setup, each observable A can be represented by a vector $\mathbf{A}$, such that $\langle A \rangle = \mathbf{A} \cdot \mathbf{p} = \sum_i A_i p_i$. After a measurement, the probability distribution is evolved to $\mathbf{p}'$ according to the above rule, and the expectation value of A changes to $\langle A \rangle = \mathbf{A} \cdot \mathbf{p}' = \mathbf{A} \cdot M\mathbf{p} = \sum_{ij} A_i m_{ij} p_j$. This is the most clear-cut way in order to model random influences leading to compatibility violations due to the measurement.

The expectation value of a product of observables then straight-forwardly evaluates to $\langle AB \rangle = \sum_{ij} A_i B_j p_{ij}$, where p_{ij} is the joint probability distribution of the observables. The extension to a product of three or more observables proceeds in an obvious way, i.e. $\langle ABC \rangle = \sum_{ijk} A_i B_j C_k p_{ijk}$.

4.2.1 Violating KS Inequalities with Probabilistic Evolutions

In order to study the behaviour of the CHSH-inequality under random hidden-variable state changes, it is useful to introduce the quantity $K_{ij} = A_i B_j + B_i C_j + C_i D_j - D_i A_j$, where A_i denotes the value of the observable A given the hidden variable state λ_i. Thus, any K_{ij} just gives the value of χ_{CHSH} given that the hidden-variable evolution was $\lambda_i \to \lambda_j$. For this quantity, the following holds:

Proposition 4.1. *For any Kochen-Specker inequality, there exists* $K = (K_{i_1 i_2 \ldots i_n})$ *such that* $\langle \chi_{KS} \rangle \leq K^{\max}$.

Proof. Consider that each expectation value $\langle AB \rangle$ can be written as $\sum_{ij} A_i B_j p_{ij}$, and thus

$$\langle \chi_{\text{CHSH}} \rangle = \sum_{ij} K_{ij} p_{ij} \leq K^{\max} \sum_{ij} p_{ij} = K^{\max}. \qquad (4.5)$$

This generalizes immediately to cases with more observables. □

All possible values for K_{ij} in the case of the CHSH-inequality are shown in A.1.

This shows straightforwardly that a simple, deterministic evolution suffices to maximally violate the CHSH inequality: $K_{0,8} = 4$, since $\lambda_0 = (++++)$ and $\lambda_8 = (-+++)$, and thus, $\langle AB \rangle = \langle BC \rangle = \langle CD \rangle = +1$, while $\langle DA \rangle = -1$. Via probabilistic mixtures of such evolutions, it is possible to equal any value between 2 and 4; for instance, if

$p_{0,0} = 2 - \sqrt{2}$ and $p_{0,8} = \sqrt{2} - 1$, i.e. if the initially prepared state λ_0 remains undisturbed with a probability of ca. 58.6%, and flips to λ_8 with a probability of 41.4%, then the expected value $\langle \chi_{\mathrm{CHSH}} \rangle$ equals the Tsirelson bound, $2\sqrt{2}$. This already demonstrates well the spirit of the model: despite being explicitly noncontextual, a contextuality test would yield a positive result simply because of the compatibility violation effected by the hidden variable state transition.

Let us now investigate how models of the presented kind manage to produce violations of Kochen-Specker inequalities. First, consider the case in which we have perfect control over the hidden-variable state we prepare. Then, the distribution over the hidden variables is $p_i = 1$ for some i, i.e. we are with certainty in the state λ_i. We can then prove the following:

Proposition 4.2. *For any hidden-variable state λ_j, there exists a deterministic evolution such that $\langle \chi_{\mathrm{CHSH}} \rangle = 4$.*

Proof. Let λ_j be the state of the hidden variables. Then, for any observable X, $\langle X \rangle = \sum_i X_i p_i$. Since $X_i p_i = 0$ for all $i \neq j$ and $p_j = 1$, $\langle X \rangle = X_j$. Any deterministic channel now takes the distribution $\mathbf{p}$ to some $\mathbf{p}'$ with the property that $p'_k = 1$ for some k, i.e. afterwards, the system is definitely in the hidden-variable state λ_k. Hence, the expectation value of a product is equal to $\langle XY \rangle = X_j Y_k$. In order to maximally violate the CHSH inequality, we now only need for all of $A_j B_k$, $B_j C_k$, and $C_j D_k$ to equal 1, while $D_j A_k$ must equal -1. But such a state can always be found: it is uniquely the state λ_k for which $B_k = A_j$, $C_k = B_j$, $D_k = C_j$ and $A_k \neq D_j$. □

Another important consideration is what happens in the case of a channel that only leads to very weak compatibility violations, i.e. in which the state is only changed with a certain, small, probability p; this models the realistic case of small measurement-induced disturbances. In this case, we have:

Proposition 4.3. *For any hidden-variable state λ_j, there exists a Markov channel inducing probabilistic state changes such that the*

CHSH-inequality is violated for arbitrarily small transition probabilities.

Proof. As before, since we start out in a definite state, $\langle X \rangle = X_j$ for the observable measured first in a sequence. As was shown above, there exists always a state λ_k such that $\langle \chi_{\text{CHSH}} \rangle$ is maximal for any λ_j. Thus, we shall assume that the system always either transitions to this state with probability p, or remains in the same state with probability $1 - p$. Then, any expectation value of a sequence of two observables can be written as $\langle XY \rangle = X_j(pY_j + (1 - p)Y_k)$. For the sequences AB, BC, and CD, this does not change anything, since $A_jB_j = A_jB_k$, $B_jC_j = B_jC_k$, and $C_jD_j = C_jD_k$. However, for the sequence DA, we get $\langle DA \rangle = p(-1) + (1 - p)(+1) = 1 - 2p$, since $D_jA_j = -D_jA_k$. Thus, the value of $\langle \chi_{\text{CHSH}} \rangle$ evaluates to

$$\langle \chi_{\text{CHSH}} \rangle = \langle AB \rangle + \langle BC \rangle + \langle CD \rangle - \langle DA \rangle = 2 + 2p, \qquad (4.6)$$

i.e. $\langle \chi_{\text{CHSH}} \rangle > 2$ for all $p > 0$. $\qquad\qquad \square$

Alternatively, this follows immediately from the fact that there always exists a deterministic evolution maximally violating the CHSH inequality (see 4.2), and an evolution yielding $\langle \chi_{\text{CHSH}} \rangle = 2$, the convex combination of which then leads to $\langle \chi_{\text{CHSH}} \rangle = 2 + 2p$.

The noncontextual nature of this model can be made more explicit by considering the representation 4.2 of the hidden-variable states. Take, for instance, the hidden- variable state $\lambda_0 = (+ + + +)$. Then, in the notation introduced in 4.2, the Markov channel can be written as:

$$M = \begin{pmatrix} 1 - p & p \\ p & 1 - p \end{pmatrix} \otimes \begin{pmatrix} 1 & 0 \\ 0 & 1 \end{pmatrix}^{\otimes 3} \qquad (4.7)$$

In this representation, it is obvious that the channel acts exclusively on the observable A, flipping its value with probabilitiy p, regardless of the other observables (whose values it leaves invariant).

Thus, the CHSH inequality can be violated by a model such as the above for arbitrarily small violations of compatibility, which are unavoidable in any real experiment.

Now, turn to the other extreme: if we assume absolute uncertainty about the preparation of the hidden variables, i.e. $p_i = \frac{1}{16}$ for all i, we obtain:

Proposition 4.4. *In the case of absolute preparation uncertainty, i.e.* $p(\lambda_i) = \frac{1}{16}$ *for all* λ_i, *there exists a unique evolution such that* $\langle \chi_{CHSH} \rangle = 4$.

Proof. This result follows directly from proposition 4.2, which asserts that there always is a maximally CHSH-violating evolution for any state. It then suffices to take the convex combination of all these evolutions, i.e. choose p_{ij} such that $p_{ij} = \frac{1}{16}$ if $K_{ij} = 4$ and $p_{ij} = 0$ otherwise. □

This directly entails that for all possible initial probability distributions, one can always find evolutions such that the CHSH inequality is violated to any desired degree.

4.2.2 Violating Extended KS-Inequalities

Recall that it was proven in [44] that for each hidden-variable dynamics obeying Assumption 2.1 (cumulative noise), the extended CHSH inequality

$$\langle \chi_{CHSH} \rangle - 2p^{err}[B_1 A_2 B_3] - 2p^{err}[C_1 B_2 C_3]$$
$$- 2p^{err}[D_1 C_2 D_3] - 2p^{err}[A_1 D_2 A_3] \leq 2 \qquad (4.8)$$

holds. In order to evaluate this in the present context, we first show the following:

Proposition 4.5. *Assumption 2.1 does not hold for random evolutions in general.*

In order to prove this, we first establish the following lemma:

Lemma 4.6. *Any Markov matrix M such that $m_{ij} = m_{ji}$, and $m_{ij} = 1$ for exactly one value of j, i.e. which is both symmetric and deterministic, squares to the unit matrix.*

Proof. The elements of the square of M, $M_{ij}^2 = \sum_k m_{ik} m_{kj}$, are non-vanishing if, and only if, $i = j$, since $m_{ik} = m_{ki}$ by symmetry, and $m_{kj} = 0$ for all $j \neq i$ because of the Markovianicity condition 4.4 and the channel's determinism. Because of the latter, every non-zero value also must be equal to 1, and thus, $M^2 = \mathbb{1}$.   □

With this, we can now prove the theorem:

Proof. Consider a symmetric, deterministic Markov matrix M_{sym}. By lemma 4.6, $M_{\mathrm{sym}}^2 = \mathbb{1}$. Thus, evolving the original probability distribution twice returns it to its original state. But this means that $p[(B_1^+, B_3^- | B_1 A_2 B_3)] = 0$, while $p[(B_1^+ | B_1) \text{ and } (B_2^- | A_1 B_2)]$ is arbitrary (equal to 1, for instance, if the state is changed after *every* measurement); thus, $p[(B_1^+ | B_1) \text{ and } (B_2^- | A_1 B_2)] \not\leq p[(B_1^+, B_3^- | B_1 A_2 B_3)]$, meaning that the disturbance is not cumulative.   □

Thus, not all possible dynamics for the hidden variables obey assumption 2.1. Therefore, the extension of Kochen-Specker inequalities by error terms does not necessarily completely solve the problem of compatibility: if some evolution exists such that the disturbance is non-cumulative, inequalities such as 2.22 or 4.8 may be violated by non-contextual models. Indeed, we can establish the following:

Proposition 4.7. *For any hidden-variable state λ_j, there exists a deterministic evolution maximally violating the extended CHSH-inequality and the assumption of cumulative noise (2.1).*

Proof. Proposition 4.2 assures us that we can always find a Markov channel that violates the CHSH inequality using a deterministic model, and proposition 4.5 establishes that if the model is symmetric, assumption 2.1 is violated, and in fact, all terms of the form $p[(B_1^+, B_3^- | B_1 A_2 B_3)]$, and thus, all error terms, vanish; hence, it suffices to be able to violate the original CHSH inequality, and it remains only to be shown that one can do so using a symmetric Markov matrix. But this is necessarily the case: the model capable of violating the CHSH inequality was such that it always flipped the hidden-variable according to the rules given in the proof of proposition 4.2, i.e. for instance from $\lambda_3 = (+ + - -)$ to $\lambda_1 = (+ + + -)$; thus, applying the channel again simply undoes the flip, and restores the original state. $\qquad\square$

Therefore, the same simple kind of models that are capable of violating the CHSH inequalities are also capable of violating the extended inequalities of ref. [44]. However, while they cannot perfectly rule out these models, the extended inequalities can be used to put a bound on the strength of the violation of compatibility needed:

Proposition 4.8. *For any hidden-variable state λ_j, if p represents the probability that a state transition occurs, a probabilistic Markov model that violates the extended CHSH inequality needs p to exceed $\frac{7}{8}$.*

Proof. The probability that the value of a certain observable changes is equal to the probability that the state changes, given that the state changes to one in which the observable has a different value. In order to show a violation of the assumption, it is enough to consider one measurement sequence, e.g. AB; thus, the only error probabilities are $p[(A_1^+|A_1)$ and $(A_2^-|D_1A_2)] = p[(A_1^-|A_1)$ and $(A_2^+|D_1A_2)] = p$. In order to violate assumption 2.1, we need that, for instance,

$$p[(A_1^+|A_1) \text{ and } (A_2^-|D_1A_2)] > p[(A_1^+, A_3^-|A_1D_2A_3)]. \qquad (4.9)$$

Since we need the symmetry of the transition matrix in order to violate the assumption at all, we know that this probability must be equal to the probability of remaining in the state λ_k after the second transition, i.e. of not transitioning back to the original state λ_j. Since the transition $\lambda_j \to \lambda_k$ occurs with a probability of p, and the system then remains in the state λ_k with probability $(1 - p)$, the probability for the evolution $\lambda_j \to \lambda_k \to \lambda_k$ is equal to $p - p^2$. Thus, in order to violate assumption 2.1, it is necessary that

$$p > p - p^2. \qquad (4.10)$$

This is of course the case for all $p > 0$. Thus, any symmetric probabilistic Markov model violates assumption 2.1.

It now remains to establish when such a model violates equation 2.22. As was shown in the proof of proposition 4.3, for a probabilistic evolution, $\langle \chi_{\text{CHSH}} \rangle = 2 + 2p$. In order to establish a lower bound for the probability, we need to examine the worst case, in which all observables change their value. This is the case, for instance, for the evolution $(-+-+) \to (+-+-)$. Then, all error terms contribute, and $p^{\text{err}}[B_1 A_2 B_3] = p^{\text{err}}[C_1 B_2 C_3] = p^{\text{err}}[D_1 C_2 D_3] = p^{\text{err}}[A_1 D_2 A_3] = 2(p - p^2)$. Thus, the necessary condition for this violation is

$$2 + 2p > 2 + 16(p - p^2), \qquad (4.11)$$

which yields $p > \frac{7}{8}$. $\qquad\qquad\qquad\qquad\qquad\qquad\qquad\qquad\qquad$ □

We have seen in this section that simple, random evolutions of the hidden variables can lead to violations of Kochen-Specker inequalities even if they are perfectly noncontextual: the measurement context does not factor into the Markov channel, as it is applied uniformly after each measurement. This surprising feature has its origin in the fact that violations of compatibility, which is an essential assumption in the Kochen-Specker theorem but hard to guarantee in reality, may reinforce one another in a conspiratorial way to lead to experimental results that are seemingly at variance with classical predictions. Past attempts to explicitly address this issue have been shown to be not fully adequate, though able to significantly reduce the range of possible models.

One may, of course, call into question the reasonability of the models used here. It is clear that the simplest versions can, in principle, be detected and excluded easily: a model that deterministically changes the value of one observable upon measurement may be defeated by simply measuring the same observable twice and noting the consistent discrepancy; many models that introduce a random evolution, such as the one presented above to obtain the Tsirelson bound, still produce far higher error rates than experimentally observed. However, these admittedly crude models can be refined and made more realistic in various ways: uncertainty about the prepared state together with probabilistic state changes may conspire to make the compatibility violations significantly harder to detect, and as has been shown above, models exist that can foil methods proposed previously to correct for just such deviations.

Furthermore, the problem is not so much one of practice as it is one of principle: though they may appear increasingly baroque, models can be found that agree with any measurement record and which nevertheless ought not be called contextual. To remedy this situation would certainly be of great interest.

Chapter 5

Noncontextual Evolution

THE last chapter showed, with explicit models, that it is possible to violate Kochen-Specker inequalities using noncontextual dynamics. This seems to once more dampen the hopes of a definitive experimental test of the Kochen-Specker theorem. Nevertheless, as we aim to show in this chapter, the contextuality of quantum mechanics runs deeper still, and indeed deeper than the original Kochen-Specker theorem establishes. Thus, we extend the notion of contextuality to apply not merely to the state, but to the entire evolution of a system, i.e. the succession of hidden variable states $\lambda_i \to \lambda_j \to \lambda_k \to \ldots$ it traverses. This notion of *noncontextual evolution* does not require the strict compatibility of jointly measured observables. Then, we show that existing Kochen-Specker inequalities may be reformulated such that their violation translates to a violation of noncontextual evolution, and thus, excludes a class of hidden-variable theories containing both the usual noncontextual ones (in the limit where the hidden-variable state does not undergo any evolution) and those given by the models of the previous chapter.

5.1 Noncontextually Evolving Systems

In order to make precise the notion of noncontextual evolution, we first state the following assumptions:

I. All of a system's observables have definite values at any given time (*value definiteness* (VD)).

II. It is possible to uniquely attribute to each system a sequence of
(hidden-variable) states $\lambda_i \to \lambda_j \to \lambda_k \to \ldots$ that is independent
of the measurements performed on the system (*noncontextual
evolution* (NCE)).

These are clearly weaker assumptions than those made in the Kochen-Specker definition of noncontextuality, as there exist models that
obey both I. and II., yet nevertheless violate a Kochen-Specker inequality (see theorem 4.2). However, the notion of contextuality arrived at
using these assumptions includes the Kochen-Specker notion: if we
choose the evolution such that the hidden-variable state never changes,
we get the usual formulation of noncontextuality. Thus, any experimental refutation of noncontextual evolution entails a verification of
the Kochen-Specker theorem.

Furthermore, the notion arrived at in this way does not make explicit
mention of the measurement context of compatible observables, and
is thus not vulnerable to the problem of compatibility; there is no
requirement that in a sequence of measurements $ABAC \ldots A$ the first
and last measurements of A necessarily agree (recall the discussion of
compatibility in sect. 2.6). Indeed, in general, this will not be the case
for a noncontextual evolution. The only thing that is required in this
case is that there is a fixed sequence of values that can be attributed
to any observables, corresponding to a fixed series of (hidden-variable)
states traversed by the system.

The notion of noncontextual evolution is appropriate for model-
ing systems with unknown, but essentially classical, dynamics. The
motivation for the models discussed so far is the notion that during
measurement, random influences may change the ontic state, thus
giving rise to incompatibilities between observables assumed to be
compatible. However, we are not fixed on this interpretation. Another possible view would be that the system, as characterized by its
hidden-variable state, simply evolves on its own according to rules
unknown and/or inaccessible to us. Measurement in this case could
be construed essentially classical: the objective value of some physical
property is registered, with none or only negligible interaction between

measurement device and system. The outward effect of both would be the same: states assumed 'in between' measurements are simply not registered, and thus, one can ignore them, studying instead only the 'reduced' or 'coarse-grained' evolution as depicted in figure 5.1. In this view, the probabilistic nature of the evolution may be thought of as arising simply from differences in the timing of measurements: the same sequence of measurements finds the system in different states due to its internal dynamics, even though they are essentially deterministic.

Thus, we see that the notion of a noncontextually evolving theory is a quite general one, independent of the microscopic details of a possible hidden-variable theory in the sense that only observed quantities matter, not the dynamics of how the system evolves from one state to another.

5.2 Inequalities for All Noncontextually Evolving Systems

From proposition 4.1, we know that the value of a Kochen-Specker inequality is upper bounded by the quantity K^{max}. As a brief reminder, for the CHSH-inequality, this is the maximum of $K_{ij} = A_i B_j + B_i C_j + C_i D_j - D_i A_j$, i.e. 4. This is the highest value that can be reached by models undergoing noncontextual evolution—and, as the discussion in the previous chapter shows, this limit is actually attainable.

Can an inequality be found, such that the quantum value exceeds the value attainable by noncontextual evolution? If so, experimentally verifying this violation would amount to both a conclusive test of the Kochen-Specker theorem, since observables in this scenario may exhibit arbitrary violations of compatibility, and exclude an actually even larger class of hidden variable theories, namely those exhibiting noncontextual evolution.

The strategy is thus clear: if we can find an inequality such that K^{max} is smaller than the quantum value, its experimental violation

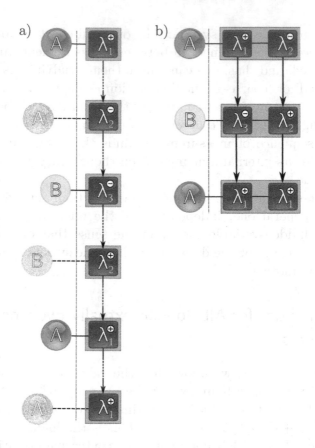

Figure 5.1: **a)** Measuring a system that undergoes a deterministic, but unknown and uncontrollable evolution: measurements executed at different times during different runs of the experiment (full resp. broken circles) may yield different results, while states of the system in between measurements have no observational consequences. **b)** The same system as in a), however, the evolution is now understood as a probabilistic superposition of two possible evolutions, leading to the same observations.

would establish that quantum mechanics cannot be explained in terms of a noncontextually evolving hidden variable theory and, by restriction to the case in which all observables are compatible, that it thus cannot be explained in terms of a (Kochen-Specker) noncontextual realistic theory. Remarkably, such inequalities can indeed be found, and are given by slight modifications of already familiar Kochen-Specker inequalities.

5.2.1 Generalizing the CHSH-Inequality

We shall start our investigations with the CHSH-inequality. Let us look first at its general form:

$$\langle \chi \rangle = \langle X^{(1)}Y^{(1)} \rangle + \langle X^{(2)}Y^{(2)} \rangle + \langle X^{(3)}Y^{(3)} \rangle - \langle X^{(4)}Y^{(4)} \rangle \quad (5.1)$$

From this, the form 2.4 follows by the identifications $X^{(1)} = Y^{(4)} = A$, $Y^{(1)} = X^{(2)} = B$, $Y^{(2)} = X^{(3)} = C$, and $Y^{(3)} = X^{(4)} = D$. The value of K_{ij} is computed as before:

$$K_{ij} = X_i^{(1)}Y_j^{(1)} + X_i^{(2)}Y_j^{(2)} + X_i^{(3)}Y_j^{(3)} - X_i^{(4)}Y_j^{(4)} \quad (5.2)$$

We can establish certain conditions that must hold for this expression to be maximized, i.e. equal to 4. Keeping in mind that each of $X_i^{(k)} = Y_i^{(k)} = \pm 1$, these are:

$$
\begin{aligned}
X_i^{(1)} &= Y_j^{(1)} \quad\quad\quad (5.3)\\
X_i^{(2)} &= Y_j^{(2)}\\
X_i^{(3)} &= Y_j^{(3)}\\
X_i^{(4)} &\neq Y_j^{(4)}
\end{aligned}
$$

Clearly, with the above assignment of observables leading to the form 2.4, these conditions can be met, since in general, for instance, $A_i \neq A_j$. Thus, the original form of the CHSH inequality is not an inequality capable of ruling out noncontextually evolving hidden variables, as indeed was explicitly demonstrated in the previous chapter. If, however, we could find an assignment such that not all conditions can be satisfied simultaneously, then $K^{\max}$ would necessarily be constrained to a lower value.

One way to do this is to ensure that the truth of the first three conditions in 5.3 implies the falsity of the fourth. We make the following choice:

$$
\begin{aligned}
X_i^{(1)} &= X_i^{(4)} \\
Y_j^{(1)} &= Y_j^{(2)} \\
X_i^{(2)} &= X_i^{(3)} \\
Y_j^{(3)} &= Y_j^{(4)}
\end{aligned}
\tag{5.4}
$$

This directly leads to the following:

Proposition 5.1. *The following version of the CHSH inequality, which we call simply* CHSH*, *given by*

$$
\langle \chi_{\text{CHSH}^*} \rangle = \langle AB \rangle + \langle CB \rangle + \langle CD \rangle - \langle AD \rangle \leq 2
\tag{5.5}
$$

holds for all noncontextually evolving systems, but is violated by quantum mechanics.

Proof. The proof follows straightforwardly by checking the conditions 5.3. Recall that, for CHSH*,

$$
K_{ij} = A_i B_j + C_i B_j + C_i D_j - A_i D_j.
\tag{5.6}
$$

Evaluating the conditions 5.3, $A_i = B_j$ and $C_i = B_j$, and thus, $A_i = C_i$. But then, since $C_i = D_j$, $A_i = D_j$, violating the fourth condition. Thus, $K_{ij} \leq 2$.

In order to show the quantum violation, it suffices to exhibit a set of observables and a state that yield a value of $\langle \chi_{\text{CHSH}^*} \rangle > 2$. We again choose a pair of qubits in the state $|\psi\rangle = \frac{1}{\sqrt{2}}(|00\rangle + |11\rangle)$, and the observables $A = \sigma_x \otimes \mathbb{1}$, $B = -\frac{1}{\sqrt{2}}\mathbb{1} \otimes (\sigma_z + \sigma_x)$, $C = \sigma_z \otimes \mathbb{1}$ and $D = \frac{1}{\sqrt{2}}\mathbb{1} \otimes (\sigma_z - \sigma_x))$, thus ensuring $\langle \chi_{\text{CHSH}^*} \rangle = 2\sqrt{2}$. However, there is a slight additional hitch that needs to be addressed: the value of $2\sqrt{2}$ implicitly assumes perfect compatibility, i.e. the state does not change between measurements. This, as has already been discussed, is unrealistic, and moreover would reintroduce the dependence on compatible observables into the scenario.

Luckily, we can here take advantage of the discussion in chapter 3, where it was shown that, under very general models for the noise introduced *at the quantum level* into the measurement, the value of $\langle \chi_{\text{CHSH}^*} \rangle$ only decreases (until eventually no violation is observable anymore). Thus, in order to demonstrate a violation of noncontextual evolution, a necessary and sufficient condition is

$$2 < \langle \chi_{\text{CHSH}^*} \rangle \leq 2\sqrt{2}. \tag{5.7}$$

$\square$

The values of K_{ij} for the CHSH* inequality are collected in A.2.

That this simple variation on the CHSH inequality should have such far-reaching consequences deserves some further discussion. Let us consider what happens during a measurement sequence. We start out with a probability distribution $p(\lambda)$, from which the first observable's value is drawn; then, the probability distribution is evolved, and the second observable's value is drawn from the evolved distribution. Since the evolution is noncontextual, it will be the same regardless of which observable is measured first; thus, whenever an observable is drawn second, it will be drawn from the same distribution. This then yields some intuition as to why noncontextually evolving systems cannot violate the inequality 5.5: since each observable always appears in the same place in each measurement sequence, its value is always drawn from the same probability distribution.

This intuitive argument applies only in the case of a product distribution, however; nevertheless, the result applies to non-product distributions as well.

5.2.2 Generalizing the Peres-Mermin Inequality

Taking a hint from the previous discussion, we immediately state the following:

Proposition 5.2. *The version of the Peres-Mermin inequality given by*

$$\langle \chi_{\mathrm{PM}^*} \rangle = \langle ABC \rangle + \langle cab \rangle + \langle \beta\gamma\alpha \rangle + \langle Aa\alpha \rangle + \langle \beta Bb \rangle - \langle c\gamma C \rangle \leq 4 \quad (5.8)$$

holds for all noncontextually evolving theories, but is violated by quantum mechanics.

Proof. Again, we must check the value of K^{max}. Let us first analyze the condition that must be met in order for a term such as $\langle ABC \rangle$ to be maximal. Interpreting the outcomes of measurements of the observables as truth values, i.e. considering the A_i, B_j and C_k to be propositions of the form 'the value of A, given the hidden variable state λ_i, is 1', and thus, to be represented by Boolean variables, the maximality of $\langle ABC \rangle$ is equivalent to the truth of the proposition:

$$P_{\mathrm{max}} = (A_i \wedge B_j \wedge C_k) \vee (A_i \wedge \neg B_j \wedge \neg C_k)$$
$$\vee (\neg A_i \wedge B_j \wedge \neg C_k) \vee (\neg A_i \wedge \neg B_j \wedge C_k), \quad (5.9)$$

where $\wedge$ denotes the logical **and**, $\vee$ denotes **or**, and $\neg$ stands for negation. This means nothing else than that either all three or only one of the outcomes must be $+1$ in order to have $\langle ABC \rangle = 1$. Using the rules of Boolean algebra, this proposition can be reduced to $A_i \oplus B_j \oplus C_k$, where $\oplus$ denotes the exclusive **or** or **xor** (i.e. addition modulo 2). We can thus compactly write the condition of maximality for the whole inequality, taking note of the fact that $\neg(x \oplus y \oplus z) = x \oplus y \oplus z \oplus 1$:

$$(A_i \oplus B_j \oplus C_k) \wedge (c_i \oplus a_j \oplus b_k) \wedge (\beta_i \oplus \gamma_j \oplus \alpha_k)$$
$$\wedge \quad (A_i \oplus a_j \oplus \alpha_k) \wedge (\beta_i \oplus B_j \oplus b_k) \wedge (c_i \oplus \gamma_j \oplus C_k \oplus 1) = 1 \quad (5.10)$$

But this proposition cannot be fulfilled. To see this, consider that $c_i \oplus a_j \oplus b_k = 1$ implies that $c_i = a_j \oplus b_k \oplus 1$, and similarly, $\gamma_k = \beta_i \oplus \alpha_k \oplus 1$ and $C_k = A_i \oplus B_j \oplus 1$. Substituting these into the condition for the maximality of the final, negative term in the inequality 5.8, which is

$$c_i \oplus \gamma_j \oplus C_k \oplus 1 = 1, \quad (5.11)$$

we get

$$\overbrace{a_j \oplus b_k \oplus 1}^{c_i} \oplus \overbrace{\beta_i \oplus \alpha_k \oplus 1}^{\gamma_k} \oplus \overbrace{A_i \oplus B_j \oplus 1}^{C_k} \overset{!}{=} 1. \tag{5.12}$$

Using $1 \oplus 1 = 0$, we can rearrange the above to

$$\beta_i \oplus B_j \oplus b_k \oplus A_i \oplus a_j \oplus \alpha_k \overset{!}{=} 1. \tag{5.13}$$

However, $\beta_i \oplus B_j \oplus b_k = 1$ and $A_i \oplus a_j \oplus \alpha_k = 1$ are just the conditions for the maximality of $\langle \beta B b \rangle$ and $\langle A a \alpha \rangle$, respectively.

But then, it immediately follows that 5.13 is a contradiction, and thus, $\langle \beta B b \rangle$, $\langle A a \alpha \rangle$, and $-\langle c \gamma C \rangle$ cannot simultaneously equal $+1$. Hence, K^{max} is constrained to remain below 4 for all noncontextually evolving theories.

The quantum violation of 5.8 follows from the same considerations as in the case of 5.5: again, any quantum noise induced tends only to drive the value of $\langle \chi_{\mathrm{PM}*} \rangle$ down; additionally, the quantum violation saturates the algebraic maximum of the inequality. Thus, a necessary and sufficient condition to prove a violation of noncontextual evolution is

$$4 < \langle \chi_{\mathrm{PM}*} \rangle \le 6. \tag{5.14}$$

$\square$

In this formulation, we can take full advantage of the fact that the proof of the Peres-Mermin inequality, as given in 2.2.2, is state-independent; thus, not only contextuality, but also the contextuality of evolution is a state-independent property of quantum mechanics—as it of course must be, since it includes the former notion.

As discussed so far, the method used to rearrange Kochen-Specker inequalities is limited to those inequalities that possess a particular permutational symmetry in their terms. However, as shown in [70], it is possible to enhance the approach to cover a wider set of cases. Consider, for example, the following inequality due to Klyachko, Can, Binicioğlu, and Shumovsky (KCBS), which is capable of being violated even for a single qutrit system [39]:

$$\langle \chi_{\mathrm{KCBS}} \rangle = \langle AB \rangle + \langle BC \rangle + \langle CD \rangle + \langle DE \rangle + \langle EA \rangle \ge -3. \tag{5.15}$$

Clearly, the attempt to rearrange it following the pattern used so far will not be successful. However, this can be remedied by including an additional term: modifying it to

$$\langle AB \rangle + \langle CB \rangle + \langle CD \rangle + \langle ED \rangle + \langle EA \rangle - \langle AA \rangle \geq -4, \qquad (5.16)$$

we again obtain an inequality valid for all noncontextually evolving systems, which is however violated by quantum mechanics. Other inequalities can be treated similarly.

5.3 An Extension of the KS-Theorem

In the previous section, we have arrived at inequalities valid for noncontextual evolutions (NCE-inequalities), which are:

   I. obeyed by any theory in which systems evolve noncontextually, i.e. which obey both VD and NCE as defined in section 5.1,

  II. violated by quantum mechanics,

 III. *independent* of the notion of compatible observables,

 IV. due to their operationally-definable nature, *immune* to the finite-precision problem, and

  V. a generalization of Kochen-Specker inequalities, in the sense that, *if* compatibility holds between the different measurements, their violation implies the Kochen-Specker theorem.

These properties furthermore entail the following:

1. Quantum mechanics cannot be completed by a theory in which the hidden variable state evolves noncontextually.

2. The impossibility of completing quantum mechanics with noncontextually evolving hidden variables can be tested experimentally.

3. Experimental verification of this impossibility implies a verification of the Kochen-Specker theorem.

The truth of 1 follows straightforwardly from I and II. For 2, as was discussed in chapter 2, the main obstacles to experimentally testing the Kochen-Specker theorem are the finite-precision problem (2.5) and the problem of compatibility (2.6). The second problem does not apply, since the notion of noncontextual evolution is independent of the notion of compatibility (III). Furthermore, as was also explicitly shown, the models implementing violations of compatibility as discussed in chapter 4 are incapable of violating inequalities 5.5 and 5.8. As argued in chapter 2, the finite-precision problem does not apply to inequalities which may be arrived at via an operational definition free of references to quantum theory (see sect. 2.5.2). Finally, 3 follows directly from the fact that the set of all noncontextual evolutions includes the trivial evolution, i.e. that in which the hidden variable state never changes.

This result deserves some further comments. First, we note that the notion of noncontextually evolving systems is a very intuitive one and applies to all classical systems: a thrown baseball, for instance, follows its parabolic trajectory independent of measurements of its height, position, or momentum (in so far as such measurements may be obtained without exercising significant influence on the ball, such as, for instance, optically, i.e. via video or photography).

Thus, noncontextual evolution formalizes the intuitive idea that in a classical theory, measurement is exclusively about ascertaining the value of a *possessed property* of a system, and it does so in a way that is weaker than the original Kochen-Specker notion: Kochen-Specker noncontextuality asserts that there is a definite association of properties to a system in a given state; noncontextual evolution merely asserts that there exists a set of states that the system may assume such that we can validly assert propositions of the form 'in the state λ_i, the value of the observable A is A_i'. This extends the notion of contextuality from one applicable within a given state, and thus, within a set of compatible measurements (as incompatible measurements will change the state according to the projection postulate), from one applicable to the set of possible states as a whole.

A somewhat related concept, similarly establishing the impossibility of mimicking the time evolution of quantum systems classically, was introduced by Leggett and Garg in their 1985 article "Quantum Mechanics versus Macroscopic Realism: Is the Flux There when Nobody Looks?" [71]. They define a notion of *macroscopic realism* by means of the following two postulates:

I. A macroscopic system with two or more macroscopically distinct states available to it will at all times *be* in one or the other of these states (*macroscopic realism* (MR)).

II. It is possible, in principle, to determine the state of the system with arbitrarily small perturbation on its subsequent dynamics (*noninvasive measurement* (NM)).

From these notions, they derive inequalities that any theory obeying MR and NM must satisfy, but which are violated in quantum mechanics. One such inequality is the following:

$$C_{12} + C_{23} + C_{34} - C_{14} \leq 2 \qquad (5.17)$$

Here, a symbol like C_{ij} denotes the correlation between measurements on a system undertaken at times t_i and t_j. The similarity to the CHSH* inequality is obvious if we identify $C_{12} = \langle A_1 B_2 \rangle$, $C_{23} = \langle C_3 B_2 \rangle$, $C_{34} = \langle C_3 D_4 \rangle$ and $C_{14} = \langle A_1 D_4 \rangle$, where the symbol B_2 for instance denotes the observable B measured at time t_2.

Despite this similarity, however, there are important differences. The most obvious one lies with the fact that Leggett and Garg consider only a single measurement, repeated at different times: if the system then does not undergo any dynamics in between measurements—i.e. if its state stays the same—, then quantum mechanics predicts just a repetition of the same outcome with each new measurement, and thus, perfect correlation. But then, $C_{12} + C_{23} + C_{34} - C_{14} \equiv 2$. This contrasts with the fact that for 5.5, $\langle \chi_{\text{CHSH}^*} \rangle = 2\sqrt{2}$ in the case of no state changes, due to our choice of observables. But this fact is necessary in order to conclude a violation of noncontextuality from an experimental violation of 5.5; thus, inequalities like 5.17—which after all were not derived with a focus towards contextuality—cannot be used for this purpose.

Furthermore, with their postulate II., they insist on the *noninvasiveness* of measurement, while the motivating assumption of the present work, to a large extent, was precisely the fact that measurements tend to be *invasive* in the sense that in realistic implementations, noise effects and environmental couplings tend to induce spontaneous state changes. Thus, while one can argue that the notions coincide in the case of what we have called deterministic evolutions (as we have indeed done, see Figure 5.1 and surrounding text), it seems more problematic to incorporate the notion of probabilistic mixtures of evolutions into their approach. Additionally, we have not been concerned with any notion of macrocopicity.

Nevertheless, the similarity of both results is certainly striking, and the new connection to contextuality and the Kochen-Specker theorem seems well worth pursuing. One straightforward consequence of this connection is the possibility to view the Peres-Mermin inequality in the form 5.8 as a kind of Leggett-Garg inequality involving the correlations between measurements at three different moments in time:

$$C_{123} + C_{456} + C_{789} + C_{159} + C_{726} - C_{483} \leq 4, \tag{5.18}$$

using the identifications $C_{123} = \langle A_1 B_2 C_3 \rangle$, $C_{456} = \langle c_4 a_5 b_6 \rangle$, $C_{789} = \langle \beta_7 \gamma_8 \alpha_9 \rangle$, and so on.

5.3.1 Contextually Evolving Hidden Variables

In order to fully grasp the notion of noncontextual evolution, it is instructive to consider systems that evolve contextually, and thus, are capable of violating the NCE-inequalities despite all observables having definite values at all times. Such models have, for instance, been proposed in the investigation of the memory cost of classically

simulating contextuality [72], or in Larsson's proposal to formulate a
contextual extension [73] of Spekkens' toy theory [74].

The model of ref. [72] is defined as an automaton, i.e. as a system
with a finite number of states and fixed rules according to which it
transitions between them, which can capture all the predictions of
the Peres-Mermin square (see sect. 2.2.2). The particular automaton
they choose has four distinct states:

$$S_1 = \begin{bmatrix} + & + & (+,2) \\ + & + & (+,3) \\ + & + & + \end{bmatrix} \tag{5.19}$$

$$S_2 = \begin{bmatrix} + & + & (+,2) \\ - & + & - \\ (-,4) & (+,3) & + \end{bmatrix} \tag{5.20}$$

$$S_3 = \begin{bmatrix} + & - & - \\ + & + & + \\ (+,1) & (-,4) & + \end{bmatrix} \tag{5.21}$$

$$S_4 = \begin{bmatrix} + & - & (-,3) \\ - & + & (-,3) \\ - & - & + \end{bmatrix} \tag{5.22}$$

The notation for these states includes the transition rules between
them: if, for instance, the third observable in the first row of the
Peres-Mermin square (corresponding to the observable C) is measured
while the system is in state S_1, it outputs the value $+1$ and then
transitions to the state S_2.

Let us investigate how the model violates the NCE-inequality 5.8.
We will assume that for every sequence of measurements, the automaton is initialized in the state S_1. We will denote the state before and
after measurement as subscripts, and the measurement outcome as
superscript on the symbol for the observable measured, in the form
$_{S_1}C^+_{S_2}$, meaning that the automaton was in the state S_1 before measurement, the measuring of C produced the outcome $+1$ and caused
the automaton to transition to the state S_2. Thus, an expectation

value such as $\langle ABC \rangle$ evaluates to $_{S_1} A_{S_1}^+ B_{S_1}^+ C_{S_2}^+ = 1$. Then, we can calculate $\langle \chi_{\mathrm{PM}*} \rangle$ as follows:

$$\langle \chi_{\mathrm{PM}*} \rangle = \underbrace{_{S_1} A_{S_1}^+ B_{S_1}^+ C_{S_2}^+}_{1} + \underbrace{_{S_1} c_{S_2}^+ a_{S_2}^+ b_{S_2}^+}_{1} + \underbrace{_{S_1} \beta_{S_1}^+ \gamma_{S_1}^+ \alpha_{S_1}^+}_{1} \qquad (5.23)$$

$$+ \; \underbrace{_{S_1} A_{S_1}^+ a_{S_1}^+ \alpha_{S_1}^+}_{1} + \underbrace{_{S_1} \beta_{S_1}^+ B_{S_1}^+ b_{S_1}^+}_{1} - \underbrace{_{S_1} c_{S_3}^+ \gamma_{S_3}^+ C_{S_1}^-}_{-1} = 6$$

Thus, the automaton correctly reproduces the quantum value $\langle \chi_{\mathrm{PM}*} \rangle = 6$. This means that the automaton evolves in a contextual way, as can also be seen by the fact that its evolution cannot be depicted by a diagram of the form of 4.2; the branching does not occur in a probabilistic way, but rather, deterministically based on the measurements performed on the system. Two example evolutions are given in Figure 5.2.

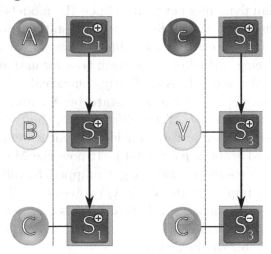

Figure 5.2: Two evolutions of the automaton from ref. [72]; the state change after the measurement of C in the first evolution is not indicated.

Thus, systems of this kind may, in principle at least, underly the dynamics of contextually evolving systems. However, the transition rules 5.19–5.22 are rather ad hoc and unnatural (of course, it was not the purpose of ref. [72] to find a natural or realistic model). Hence, it would be interesting to consider a model whose rules might be more well-motivated.

Such models are given by close 'cousins' of the models discussed in chapter 4. Again, we will focus only on their capacity to violate NCE-inequalities, neglecting for the moment the question of how to fashion them into realistic models for quantum mechanics as a whole.

The key to these models is to restrict attention to only the part of the state relevant to a given measurement; i.e. if the state is $\lambda_4 = (+-++)$, and we are considering the measurement sequence AB, the relevant part is simply the tuple $(+-)$, which we name, in analogy with the naming convention we used for the λ_i, μ_i. Thus, if we consider the CHSH* inequality, the relevant state space is $\{\mu_0, \mu_1, \mu_2, \mu_3\} = \{(++), (+-), (-+), (--)\}$. On this reduced state space, we let the same Markovian dynamics act that defined the models in chapter 4. This has the effect that, since different reduced states evolve differently, and the same full state reduces in different ways depending on the measurement being carried out, the state transformation depends on the measurement—and is thus explicitly contextual.

To give an example, let the full state be λ_4, and consider the measurement sequence AB. As defined above, the reduced state is then $\mu_1 = (+-)$. However, if we consider the measurement sequence CD, the reduced state is $\mu_0 = (++)$. If now the Markov channel is such that the states μ_1 and μ_0 get mapped to different states, this means that implicitly the state λ_4 evolves in a different way if the measurement sequence is AB than if it is CD; we thus have a contextually evolving system.

We can now show the following:

Proposition 5.3. *There exist models of the above-discussed kind capable of violating NCE-inequalities.*

Proof. The proof follows by exhibiting a concrete example: the model implemented by the Markov matrix

$$M_\mu = \begin{pmatrix} 1 & 0 & 0 & 1 \\ 0 & 0 & 1 & 0 \\ 0 & 1 & 0 & 0 \\ 0 & 0 & 0 & 0 \end{pmatrix} \tag{5.24}$$

violates the CHSH* inequality 5.5 maximally *if* the hidden variables start out in the state $\lambda_9 = (-++-)$. The proof follows by direct calculation of the expectation value $\langle \chi_{\text{CHSH}*} \rangle$. As a reminder, the CHSH* inequality is

$$\langle \chi_{\text{CHSH}*} \rangle = \langle AB \rangle + \langle CB \rangle + \langle CD \rangle - \langle AD \rangle \le 2. \tag{5.25}$$

Since the hidden variable state in the beginning is λ_9, the reduced states relevant for the calculation are, in order:

- $\mu_2 = (-+)$ for the measurement of AB,

- $\mu_0 = (++)$ for the measurement of CB,

- $\mu_1 = (+-)$ for the measurement of CD, and

- $\mu_3 = (--)$ for the measurement of AD.

The stochastic matrix 5.24 tells us how the state is changed during each measurement: a one in the position i, j indicates that the state μ_i is always changed to μ_j after the first observable is measured. Using notation analogous to 5.23, we get the expectation values:

$$\langle \chi_{\text{CHSH}*} \rangle = \underbrace{{}_{\mu_2} A^-_{\mu_1} B^-_{\mu_3}}_{1} + \underbrace{{}_{\mu_0} C^+_{\mu_0} B^+_{\mu_0}}_{1} + \underbrace{{}_{\mu_1} C^+_{\mu_3} D^+_{\mu_0}}_{1} - \underbrace{{}_{\mu_3} A^-_{\mu_0} D^+_{\mu_0}}_{-1} = 4$$

$$\tag{5.26}$$

The model thus violates the $CHSH^*$ inequality maximally, yielding $\langle \chi_{\text{CHSH}*} \rangle = 4$. □

This model has some peculiar properties that are worth investigating. First, let us generalize to a case in which the state transition only occurs probabilistically, i.e. to a model given by the Markov matrix $M_\mu(p)$:

$$M_\mu(p) = \begin{pmatrix} 1 & 0 & 0 & p \\ 0 & 1-p & p & 0 \\ 0 & p & 1-p & 0 \\ 0 & 0 & 0 & 1-p \end{pmatrix} \tag{5.27}$$

Here, p is the probability that a given state is changed, and thus, $1 - p$ is the probability that it remains unchanged.

We can immediately establish an analogous result to proposition 4.3:

Proposition 5.4. *The model implemented by the stochastic matrix 5.27 violates the $CHSH^*$-inequality for any value of $p > \frac{2}{3}$.*

Proof. The proof follows via direct calculation. The expectation value of a product of two observables, say, A and B can be written as $\langle AB \rangle = p \left({}_{\mu_2} A^-_{\mu_1} B^- \right) + (1-p) \left({}_{\mu_2} A^-_{\mu_2} B^+ \right) = p(+1) + (1-p)(-1) = -1 + 2p$; similarly, $\langle CB \rangle = 1$, $\langle CD \rangle = -1 + 2p$ and $\langle AD \rangle = 1 - 2p$. Hence,

$$\langle \chi_{\text{CHSH}^*} \rangle = \langle AB \rangle + \langle CB \rangle + \langle CD \rangle - \langle AD \rangle = -2 + 6p \tag{5.28}$$

Thus, $\langle \chi_{\text{CHSH}^*} \rangle > 2$ requires $p > \frac{2}{3}$. □

This might seem to immediately disqualify such a model due to unacceptably large violations of compatibility. However, the most immediate, naive test of compatibility, checking the requirement that $[X, Y] = 0$ for all observables X and Y in the same context, will not work as expected:

Proposition 5.5. *Despite its obvious violations of compatibility, the model as given by 5.27 is pseudo-compatible, i.e. $\langle [X, Y] \rangle = 0$ for all observables within a context.*

Proof. We will establish this by explicitly checking the commutators. For this, it suffices to check whether $\langle BA \rangle$, $\langle BC \rangle$, $\langle DC \rangle$ and $\langle DA \rangle$ agree with their counterparts calculated in the proof of proposition 5.4. This yields $\langle BA \rangle = p \left({}_{\mu_1} B^+_{\mu_2} A^+ \right) + (1-p) \left({}_{\mu_1} B^+_{\mu_1} A^- \right) = p(+1) + (1-p)(-1) = -1 + 2p$, $\langle BC \rangle = 1$, $\langle DC \rangle = -1 + 2p$ and $\langle DA \rangle = 1 - 2p$. Hence, the expectation value of all commutators vanishes.   □

This shows that, in order to establish compatibility, it is not enough to require the commutator to vanish, contrary to what is sometimes claimed in the literature (e.g. [32, 49]).

Chapter 6
Conclusion

THE main topic of this book has been to examine the experimental testability of the Kochen-Specker theorem, with a focus towards the problem of compatibility. The approach we have taken towards the Kochen-Specker theorem, and to no-go theorems in general, was motivated by the debate about the possibility of a completion of quantum mechanics by hidden variables, such that the resulting theory yields definite and unambiguous predictions for all observable quantities.

As has already been remarked upon, it is not possible to rule out hidden variable theories in general: a perfectly adequate hidden variable theory can, for instance, always be given by the observed probability distributions in any experiment themselves. Thus, the best one can do is to put limits on possible completions, and then consider whether what is left really can give rise to a theory whose implications are any less radical than those of quantum mechanics. To this end, it is important to collect experimental evidence for the validity of no-go theorems[1], as has been already done with great success in the case of Bell's theorem, in order to establish which of a pair (or set) of mutually exclusive possibilities is in fact realized in nature.

Thus, in chapter 3, we have first undertaken an investigation into the behaviour of certain noncontextuality inequalities under noisy

---

[1] Of course, there is not really a way to test a theorem experimentally; in as much as it is a piece of pure mathematics, it is exactly as good and valid as the assumptions and reasoning that went into its derivation. Rather, it is the experimental consequences of a theorem that are being tested, and language of the kind used above should be understood as shorthand for this.

measurements, in order to quantify the degree of violation to be expected in real experiments.

Under a wide variety of noise models (depolarizing noise (3.3), bit-flip errors (3.4), amplitude damping (3.5), and phase damping (3.6)), it was found that the quantum violation decreased steadily, but not too fast in order to expect to be able to observe violations of noncontextuality of quantum mechanic origin in real experiments. Furthermore, we analyzed the extended Kochen-Specker inequalities of ref. [44], for which the bounds were found to be significantly tighter, but still not unmatchable.

Then, in chapter 4, we introduced classical models designed to produce violations of compatibility in order to give the appearance of violations of noncontextuality. The operational motivation of these models was rooted in the measurement process: since we have no control over any possible hidden-variable state underlying the quantum state of a system, interactions between the measurement device and the measured system may cause unpredictable and effectively random hidden-variable state transitions. Such transitions then show up in the form of violations of compatibility. As was shown in proposition 4.2, for any possible hidden-variable state, there exists an evolution such that the CHSH inequality is maximally violated, despite this evolution being independent of the measurement context and thus in this sense noncontextual.

Furthermore, we have subjected the extended CHSH inequality of ref. [44] to the same analysis, and found that, while it is not as easily violated as the regular CHSH inequality, it—and, by extension, similar extensions of other inequalities—nevertheless cannot exclude these models.

Finally, chapter 5 saw the formulation of a first attempt to overcome the problem of compatibility. To this end, we formulated a notion of noncontextuality which we termed noncontextual evolution 5.1, defined by two postulates:

I. All of a system's observables have definite values at any given time (*value definiteness* (VD)).

II. It is possible to uniquely attribute to each system a sequence of (hidden-variable) states $\lambda_i \rightarrow \lambda_j \rightarrow \lambda_k \rightarrow \ldots$ that is independent of the measurements performed on the system (*noncontextual evolution* (NCE)).

As shown in chapter 4, systems obeying I. and II. may exhibit violations of Kochen-Specker inequalities. However, the notion of noncontextual evolutions led us to new inequalities—actually, reformulations of inequalities 2.4 and 2.18 in which attention was paid to the ordering of observables—, that were shown to be obeyed by all noncontextually evolving systems, yet are nevertheless violated by quantum mechanics, the contextuality of which thus was shown to run even deeper than the Kochen-Specker notion.

The advantage of noncontextual evolution is that it does not depend on the notion of compatibility: while it is not possible to derive inequalities such as 2.4 and 2.18 without assuming the co-measurability (and hence, compatibility) of the observables, and thus, the experimental violation of these inequalities does not say anything about the Kochen-Specker notion of contextuality, inequalities like the one we called CHSH*, given by

$$\langle \chi_{\text{CHSH}^*} \rangle = \langle AB \rangle + \langle CB \rangle + \langle CD \rangle - \langle AD \rangle \overset{HV}{\leq} 2, \qquad (6.1)$$

hold even if there is no perfect compatibility between the observables. The only requirement is for there to be a definite evolution (or probabilistic combination of such evolutions) $\lambda_i \rightarrow \lambda_j \rightarrow \ldots$ that applies independently of the measurements being carried out on the system, i.e. the context. Thus, experimental observation of the violation of these inequalities does unambiguously indicate that quantum mechanics cannot be replaced by a noncontextually evolving theory, since, due to the investigations of chapter 3, we know that even under imperfect measurements, quantum mechanics violates such inequalities. Additionally, since any system for which the Kochen-Specker notion of noncontextuality applies is a noncontextually evolving system—its evolution being the trivial one in which the hidden variable state never changes—, this experimental violation also implies that quantum mechanics cannot be supplanted by Kochen-Specker noncontextual hidden variables.

In fact, as commented upon in [70], an experiment testing whether nature obeys the constraints of noncontextual evolution has effectively already been carried out by Kirchmair *et al.* in [45]. There, they measure all possible combinations of compatible observables in the Peres-Mermin square (2.1), including those necessary for testing the inequality 5.8. With this data, the observed value is $\langle \chi_{PM^*} \rangle_{QM} = 5.35(4)$, comfortably violating the noncontextual-evolution bound $\langle \chi_{PM^*} \rangle_{NCE} = 4$.

An interesting side result to this discussion has been the realization of a connection between our NCE-inequalities and Leggett-Garg inequalities, and thus, between noncontextuality and macroscopic realism. In a sense, an inequality such as 6.1 above may be considered as both a Leggett-Garg and Kochen-Specker inequality: it rules out macroscopic realism for nontrivial evolutions in between measuremens, and noncontextuality in the case of no state changes (which the usual Leggett-Garg inequalities do not). Thus, the notion of noncontextual evolution may be considered to provide a unified background to both.

We then seem to have arrived at a possible answer to the problem of compatibility, and thus, at a route towards subjecting the Kochen-Specker theorem to the same rigorous empirical testing that Bell's theorem already has withstood so magnificently.

We hope that this present work will thus aid in the program to put the Kochen-Specker theorem on the same firm experimental footing that Bell's theorem already rests upon, possibly opening up ways for quantum contextuality to play a role in the foundations and applications of quantum mechanics similarly to that enjoyed by nonlocality. Some first steps in this direction have already been undertaken: contextuality has been implicated to play a role in measurement-based quantum computation 2.1, it has been used to guarantee the security of quantum key distribution protocols [76, 77], and its role in axiomatizations of physical theories has been investigated [25].

A Some Tables of Values

A.1 Possible Values for the CHSH-Inequality

For the CHSH-inequality, the quantity $K_{ij} = A_i B_j + B_i C_j + C_i D_j - D_i A_j$, for all combinations, evaluates to (negative values are marked by an overbar, e.g. $\bar{2} \equiv -2$):

$$
\begin{bmatrix}
2 & 0 & 0 & 0 & 4 & \bar{2} & \bar{2} & 2 & \bar{2} & 2 & 2 & \bar{4} & 0 & 0 & 0 & \bar{2} \\
4 & 2 & 2 & 2 & 2 & 0 & 0 & 0 & 0 & 0 & 0 & \bar{2} & \bar{2} & \bar{2} & \bar{2} & \bar{4} \\
0 & 2 & \bar{2} & \bar{2} & 2 & 0 & 0 & 4 & \bar{4} & 0 & 0 & \bar{2} & 2 & 2 & \bar{2} & 0 \\
0 & \bar{2} & 2 & \bar{2} & 2 & 0 & \bar{4} & 0 & 0 & 4 & 0 & \bar{2} & 2 & \bar{2} & 2 & 0 \\
0 & \bar{2} & \bar{2} & 2 & 2 & \bar{4} & 0 & 0 & 0 & 0 & 4 & \bar{2} & \bar{2} & 2 & 2 & 0 \\
2 & 4 & 0 & 0 & 0 & 2 & 2 & 2 & \bar{2} & \bar{2} & \bar{2} & 0 & 0 & 0 & \bar{4} & \bar{2} \\
2 & 0 & 4 & 0 & 0 & 2 & \bar{2} & \bar{2} & 2 & 2 & \bar{2} & 0 & 0 & \bar{4} & 0 & \bar{2} \\
2 & 0 & 0 & 4 & 0 & \bar{2} & 2 & \bar{2} & 2 & \bar{2} & 2 & 0 & \bar{4} & 0 & 0 & \bar{2} \\
\bar{2} & 0 & 0 & \bar{4} & 0 & 2 & \bar{2} & 2 & \bar{2} & 2 & \bar{2} & 0 & 4 & 0 & 0 & 2 \\
\bar{2} & 0 & \bar{4} & 0 & 0 & \bar{2} & 2 & 2 & \bar{2} & \bar{2} & 2 & 0 & 0 & 4 & 0 & 2 \\
\bar{2} & \bar{4} & 0 & 0 & 0 & \bar{2} & \bar{2} & \bar{2} & 2 & 2 & 2 & 0 & 0 & 0 & 4 & 2 \\
0 & 2 & 2 & \bar{2} & \bar{2} & 4 & 0 & 0 & 0 & 0 & \bar{4} & 2 & 2 & \bar{2} & \bar{2} & 0 \\
0 & 2 & \bar{2} & 2 & \bar{2} & 0 & 4 & 0 & 0 & \bar{4} & 0 & 2 & \bar{2} & 2 & \bar{2} & 0 \\
0 & \bar{2} & 2 & 2 & \bar{2} & 0 & 0 & \bar{4} & 4 & 0 & 0 & 2 & \bar{2} & \bar{2} & 2 & 0 \\
\bar{4} & \bar{2} & \bar{2} & \bar{2} & \bar{2} & 0 & 0 & 0 & 0 & 0 & 0 & 2 & 2 & 2 & 2 & 4 \\
\bar{2} & 0 & 0 & 0 & \bar{4} & 2 & 2 & \bar{2} & 2 & \bar{2} & \bar{2} & 4 & 0 & 0 & 0 & 2
\end{bmatrix}
\tag{A.1}
$$

A.2 Possible Values for the CHSH*-Inequality

Analogous to the CHSH-inequality, the values of $K_{ij} = A_i B_j + C_i B_j + C_i D_j - A_i D_j$ for the CHSH*-inequality are (negative values are marked by an overbar, e.g. $\bar{2} \equiv -2$):

$$
\begin{bmatrix}
2 & 2 & 2 & \bar{2} & 2 & 2 & \bar{2} & 2 & \bar{2} & 2 & \bar{2} & \bar{2} & 2 & \bar{2} & \bar{2} & \bar{2} \\
2 & 2 & 2 & \bar{2} & 2 & 2 & \bar{2} & 2 & \bar{2} & 2 & \bar{2} & \bar{2} & 2 & \bar{2} & \bar{2} & \bar{2} \\
\bar{2} & 2 & \bar{2} & \bar{2} & \bar{2} & 2 & 2 & 2 & \bar{2} & \bar{2} & \bar{2} & 2 & 2 & 2 & \bar{2} & 2 \\
2 & 2 & 2 & \bar{2} & 2 & 2 & \bar{2} & 2 & \bar{2} & 2 & \bar{2} & \bar{2} & 2 & \bar{2} & \bar{2} & \bar{2} \\
2 & \bar{2} & 2 & 2 & 2 & \bar{2} & \bar{2} & \bar{2} & 2 & 2 & 2 & \bar{2} & \bar{2} & \bar{2} & 2 & \bar{2} \\
\bar{2} & 2 & \bar{2} & \bar{2} & \bar{2} & 2 & 2 & 2 & \bar{2} & \bar{2} & \bar{2} & 2 & 2 & 2 & \bar{2} & 2 \\
2 & 2 & 2 & \bar{2} & 2 & 2 & \bar{2} & 2 & \bar{2} & 2 & \bar{2} & \bar{2} & 2 & \bar{2} & \bar{2} & \bar{2} \\
2 & \bar{2} & 2 & 2 & 2 & \bar{2} & \bar{2} & \bar{2} & 2 & 2 & 2 & \bar{2} & \bar{2} & \bar{2} & 2 & \bar{2} \\
\bar{2} & 2 & \bar{2} & \bar{2} & \bar{2} & 2 & 2 & 2 & \bar{2} & \bar{2} & \bar{2} & 2 & 2 & 2 & \bar{2} & 2 \\
\bar{2} & \bar{2} & \bar{2} & 2 & \bar{2} & \bar{2} & 2 & \bar{2} & 2 & \bar{2} & 2 & 2 & \bar{2} & 2 & 2 & 2 \\
2 & \bar{2} & 2 & 2 & 2 & \bar{2} & \bar{2} & \bar{2} & 2 & 2 & 2 & \bar{2} & \bar{2} & \bar{2} & 2 & \bar{2} \\
\bar{2} & 2 & \bar{2} & \bar{2} & \bar{2} & 2 & 2 & 2 & \bar{2} & \bar{2} & \bar{2} & 2 & 2 & 2 & \bar{2} & 2 \\
\bar{2} & \bar{2} & \bar{2} & 2 & \bar{2} & \bar{2} & 2 & \bar{2} & 2 & \bar{2} & 2 & 2 & \bar{2} & 2 & 2 & 2 \\
2 & \bar{2} & 2 & 2 & 2 & \bar{2} & \bar{2} & \bar{2} & 2 & 2 & 2 & \bar{2} & \bar{2} & \bar{2} & 2 & \bar{2} \\
\bar{2} & \bar{2} & \bar{2} & 2 & \bar{2} & \bar{2} & 2 & \bar{2} & 2 & \bar{2} & 2 & 2 & \bar{2} & 2 & 2 & 2 \\
\bar{2} & \bar{2} & \bar{2} & 2 & \bar{2} & \bar{2} & 2 & \bar{2} & 2 & \bar{2} & 2 & 2 & \bar{2} & 2 & 2 & 2
\end{bmatrix}
\tag{A.2}
$$

Bibliography

[1] N. N. Vorobev, *Consistent families of measures and their extensions*, Theor. Probab. Appl. **7**, 147 (1962)

[2] A. Einstein, B. Podolsky, N. Rosen, *Can Quantum-Mechanical Description of Physical Reality be Considered Complete?*, Phys. Rev. **41**, 777 (1935)

[3] D. Bohm, Y. Aharonov, *Discussion of Experimental Proof for the Paradox of Einstein, Rosen, and Podolsky*, Phys. Rev. **108**, 1070 (1957)

[4] J. von Neumann, *Mathematische Grundlagen der Quantenmechanik*, Springer, Berlin (1932); English translation: *Mathematical foundations of quantum mechanics*, Princeton Univ. Press, Princeton (1955)

[5] J. S. Bell, *On The Problem of Hidden Variables in Quantum Mechanics*, Rev. Mod. Phys. **38**, 447 (1966)

[6] J. M. Jauch, C. Piron, *Can Hidden Variables be Excluded in Quantum Mechanics?*, Helv. Phys. Acta **36**, 827 (1963)

[7] J. S. Bell, *On The Einstein-Podolsky-Rosen Paradox*, Physics **1**, (1964)

[8] J. F. Clauser, M. A. Horne, A. Shimony and R. A. Holt, *Proposed experiment to test local hidden-variable theories*, Phys. Rev. Lett. **23**, 880 (1969)

[9] B. S. Tsirelson, *Quantum Generalizations of Bell's Inequality*, Lett. Math. Phys. **4**, 93 (1980)

[10] S. J. Freedman and J. F. Clauser, *Experimental test of local hidden-variable theories*, Phys. Rev. Lett. **28**, 938 (1972)

[11] A. Aspect, P. Grangier, and G. Roger, *Experimental Tests of Realistic Local Theories via Bell's Theorem*, Phys. Rev. Lett. **47**, 460 (1981)

[12] A. Aspect, P. Grangier, and G. Roger, *Experimental Realization of Einstein-Podolsky-Rosen-Bohm Gedankenexperiment: A New Violation of Bell's Inequalities*, Phys. Rev. Lett. **49**, 91 (1982)

[13] A. Aspect, J. Dalibard and G. Roger, *Experimental Test of Bell's Inequalities Using Time-Varying Analyzers*, Phys. Rev. Lett. **49**, 1804 (1982)

[14] G. Weihs, T. Jennewein, C. Simon, H. Weinfurter, and A. Zeilinger, *Violation of Bell's inequality under strict Einstein locality conditions*, Phys. Rev. Lett. **81**, 5039 (1998)

[15] S. Popescu, *Bells inequalities versus teleportation: What is non-locality?* Phys. Rev. Lett. **72**, 797 (1994)

[16] R. Horodecki, P. Horodecki, and M. Horodecki, *Violating Bell inequality by mixed spin-$\frac{1}{2}$ states: necessary and sufficient condition*, Phys. Lett. A. **200**, 340 (1995)

[17] N. Gisin, *Bell's Inequality Holds for All Non-Product States*, Phys. Lett. A **154**, 201 (1991)

[18] Sixia Yu, Qing Chen, Chengjie Zhang, C. H. Lai, and C. H. Oh, *All entangled pure states violate a single Bell's inequality*, Phys. Rev. Lett. **109**, 120402 (2012)

[19] A. M. Gleason, *Measures on the Closed Subspaces of a Hilbert Space*, J. Math. Mech. **6**, 885 (1957)

[20] E. Specker, *Die Logik nicht gleichzeitig entscheidbarer Aussagen*, Dialectica **14**, 239 (1960)

[21] M. Redhead, *Incompleteness, Nonlocality, and Realism. A Prolegomenon to the Philosophy of Quantum Mechanics*, Clarendon Press, Oxford (1987)

[22] M. Rédei, *Quantum Logic in Algebraic Approach*, Kluwer Academic Publishers, Dordrecht (2010)

[23] S. Kochen and E. Specker, *The Problem of Hidden Variables in Quantum Mechanics*, J. Math. Mech. **17**, 59 (1967)

[24] N. D. Mermin, *Hidden variables and the two theorems of John Bell*, Rev. Mod. Phys. **65** n. 3, 803 (1993)

[25] A. Cabello, S. Severini, and A. Winter, *(Non-)Contextuality of Physical Theories as an Axiom*, arXiv:1010.2163 (2010)

[26] A. Cabello, S. Filipp, H. Rauch, and Y. Hasegawa, *Proposed Experiment for Testing Quantum Contextuality with Neutrons*, Phys. Rev. Lett. **100**, 120404, (2008)

[27] I. Pitowsky, *Quantum Probability – Quantum Logic*, Springer, Berlin (1989)

[28] A. Cabello, J. Estebaranz, and G. García-Alcaine, *Bell-Kochen-Specker Theorem: A Proof with 18 vectors*, Phys. Lett. A **212**, 183 (1996)

[29] A. Peres, *Incompatible results of quantum measurements*, Phys. Lett. A **151**, 107 (1990)

[30] N. D. Mermin, *Simple unified form for the major no-hidden-variables theorems*, Phys. Rev. Lett. **65**, 3373 (1990)

[31] G. Greenstein and A. G. Zajonc, *The Quantum Challenge: Modern Research on the Foundations of Quantum Mechanics*, 2nd edition, Sudbury, Jones and Bartlett Publishers (2006)

[32] A. Cabello and G. García-Alcaine, *Proposed Experimental Test of the Bell-Kochen-Specker Theorem*, Phys. Rev. Lett. **80**, 1797 (1998)

[33] C. Simon, M. Zukowski, H. Weinfurter and A. Zeilinger, *A Feasible 'Kochen-Specker' Experiment with Single Particles*, Phys. Rev. Lett. **85**, 1783 (2002)

[34] Y.-F. Huang, C.-F. Li, Y.-S. Zhang, J.-W. Pan and G.-C. Guo, *Realization of All-or-nothing-type Kochen-Specker Experiment with Single Photons*, Phys. Rev. Lett. **90**, 250401 (2003)

[35] S. M. Roy, V. Singh, *Quantum Violation of Stochastic Non-Contextual Hidden Variable Theories*, Phys. Rev. A **49**, 3379 (1993)

[36] S. Basu, S. Bandyopadhyay, G. Kar, D. Home, *Bells inequality for a single spin-$\frac{1}{2}$ particle and quantum contextuality*, Phys. Lett. A **279**, 281 (2001)

[37] C. Simon, C. Brukner and A. Zeilinger, *Hidden-variable theorems for real experiments*, Phys. Rev. Lett. **86**, 4427 (2001)

[38] J.-Å. Larsson, *A Kochen-Specker Inequality*, Europhys. Lett., **58**(6), 799 (2002)

[39] A. A. Klyachko, M. A. Can, S. Binicioğlu, and A. S. Shumovsky, *Simple Test for Hidden Variables in Spin-1 Systems* Phys. Rev. Lett. **101**, 020403 (2008)

[40] A. Cabello, *Experimentally testable state-independent quantum contextuality*, Phys. Rev. Lett. **101**, 210401 (2008)

[41] M. Michler, H. Weinfurter, and M. Zukowski, *Experiments towards Falsification of Noncontextual Hidden Variable Theories*, Phys. Rev. Lett. **84**, 5457 (2000)

[42] Y. Hasegawa, R. Loidl, G. Badurek, M. Baron, and H. Rauch, *Quantum contextuality in a single-neutron optical experiment*, Phys. Rev. Lett. **97**, 230401 (2006)

[43] H. Bartosik,, J. Klepp,, C. Schmitzer,, S. Sponar,, A. Cabello,, H. Rauch, and Y. Hasegawa, *Experimental Test of Quantum Contextuality in Neutron Interferometry*, Phys. Rev. Lett. **103**, 040403 (2009)

[44] O. Gühne, M. Kleinmann, A. Cabello, J-Å. Larsson, G Kirchmair, F. Zähringer, R. Gerritsma, and C. F. Roos, *Compatibility and Noncontextuality for Sequential Measurements*, Phys. Rev. A **81**, 022121 (2010)

[45] G. Kirchmair, F. Zähringer, R. Gerritsma, M. Kleinmann, O. Gühne, A. Cabello, R. Blatt and C. F. Roos, *State-independent experimental test of quantum contextuality*, Nature **460**, 494 (2009)

[46] S. Goldstein et al., *Bell's theorem*, Scholarpedia, **6**(10), 8378, revision No. 91049 (2011)

[47] D. A. Meyer, *Finite Precision Measurement Nullifies the Kochen-Specker Theorem*, Phys. Rev. Lett. **83**, 3751 (1999)

[48] A. Kent, *Noncontextual Hidden Variables and Physical Measurements*, Phys. Rev. Lett., **83**, 3755 (1999)

[49] R. K. Clifton and A. Kent, *Simulating Quantum Mechanics by Non-Contextual Hidden Variables*, Proc. R. Soc. A, **456**, 2101 (2000)

[50] I. Pitowsky, *Deterministic model of spin and statistics*, Phys. Rev. D **27**, 2316 (1983)

[51] I. Pitowsky, *Quantum Mechanics and Value Definiteness*, Philos. Sci. **52**, 154 (1985)

[52] C. D. Godsil and J. Zaks, *Coloring the sphere*, University of Waterloo research report CORR 88-12 (1988)

[53] A. Cabello, *Comment on "Non-contextual Hidden Variables and Finite-Precision Measurements"*, arXiv:quant-ph/9911024v1 (1999)

[54] H. Havlicek, G. Krenn, J. Summhammer, and K. Svozil, *Coloring the rational quantum sphere and the Kochen-Specker theorem*, J. Phys. A: Math. Gen. **34**, 3071 (2001)

[55] D. M. Appleby, *Existential Contextuality and the Models of Meyer, Kent and Clifton*, Phys. Rev. A **65**, 022105 (2002)

[56] N. D. Mermin, *A Kochen-Specker Theorem for Imprecisely Specified Measurements*, arXiv:quant-ph/9912081 (1999)

[57] A. Cabello, *Finite-precision measurement does not nullify the Kochen-Specker theorem*, Phys. Rev. A, **65**, 052101 (2002)

[58] A. Cabello and J.-Å. Larsson, *Quantum contextuality for rational vectors*, Phys. Lett. A, **375**, 99 (2010)

[59] T. Breuer, *Kochen-Specker Theorem for Finite Precision Spin-One Measurements*, Phys. Rev. Lett. **88**, 240402, (2002)

[60] M. A. Nielsen, I. L. Chuang, *Quantum Computation and Quantum Information*, Cambridge University Press, Cambridge, (2000)

[61] J. Barrett and A. Kent, *Non-Contextuality, Finite Precision Measurement and the Kochen-Specker Theorem*, Stud. Hist. Philos. Modern Phys., **35**, 151 (2004)

[62] D. Bohm, *A Suggested Interpretation of the Quantum Theory in Terms of "Hidden Variables" I*, Phys. Rev. **85**, 166 (1952), D. Bohm, *A Suggested Interpretation of the Quantum Theory in Terms of "Hidden Variables" II*, Phys. Rev. **85**, 180 (1952)

[63] S. Yu, C.H. Oh, *State-independent proof of Kochen-Specker theorem with 13 rays*, Phys. Rev. Lett. **108**, 030402 (2012)

[64] A. Cabello, C. Budroni, O. Gühne, M. Kleinmann, and J.-Å. Larsson, *Tight inequalities for qutrit state-independent contextuality*, arXiv:1204.3741 [quant-ph] (2012)

[65] A. Cabello, *State-independent quantum contextuality and maximum nonlocality*, arXiv:1112.5149 [quant-ph] (2012)

[66] A. Cabello and M. Terra Cunha, *Proposal of a Two-Qutrit Contextuality Test Free of the Finite Precision and Compatibility Loopholes* Phys. Rev. Lett. **106**, 190401 (2011)

[67] K. Kraus, *States, Effects and Operations: Fundamental Notions of Quantum Theory*, Springer, Berlin (1983)

[68] N. Harrigan, R. W. Spekkens, *Einstein, incompleteness, and the epistemic view of quantum states*, Found. Phys. **40**, 125 (2010)

[69] S. P. Meyn and R. L. Tweedie, *Markov Chains and Stochastic Stability*, Springer-Verlag, London (1993)

[70] J. Szangolies, M. Kleinmann, O. Gühne, *Tests against noncontextual models with measurement disturbances*, Phys. Rev. A **87**, 050101(R) (2013)

[71] A. J. Leggett and A. Garg, *Quantum Mechanics versus Macroscopic Realism: Is the Flux There when Nobody Looks?*, Phys. Rev. Lett. **54**, 857 (1985)

[72] M. Kleinmann, O. Gühne, J. R. Portillo, J-Å. Larsson, and A. Cabello, *Memory Cost of Quantum Contextuality*, New J. Phys. **13**, 113011 (2011)

[73] J-Å. Larsson, *A Contextual Extension of Spekkens' Toy Theory*, AIP Conference Proceedings **1424**, 211 (2012)

[74] R. W. Spekkens, *In defense of the epistemic view of quantum states: a toy theory*, Phys. Rev. A **75**, 032110 (2007)

[75] R. Raussendorf, *Quantum computation, discreteness, and contextuality*, arXiv:0907.5449 [quant-ph] (2009)

[76] K. Horodecki, M. Horodecki, P. Horodecki, R. Horodecki, M. Pawłowski, and M. Bourennane, *Contextuality offers device-independent security*, arXiv:1006.0468 [quant-ph] (2010)

[77] S. Adhikari, D. Home, A. S. Majumdar, and A. K. Pan, *Quantum contextuality as security check in quantum cryptography using intraparticle entanglement*, arXiv:1101.0660 [quant-ph] (2011)